SpringerBriefs in Architectural Design and Technology

Series Editor

Thomas Schröpfer, Architecture and Sustainable Design, Singapore University of Technology and Design, Singapore, Singapore

Indexed by SCOPUS

Understanding the complex relationship between design and technology is increasingly critical to the field of Architecture. The *Springer Briefs in Architectural Design and Technology* series provides accessible and comprehensive guides for all aspects of current architectural design relating to advances in technology including material science, material technology, structure and form, environmental strategies, building performance and energy, computer simulation and modeling, digital fabrication, and advanced building processes. The series features leading international experts from academia and practice who provide in-depth knowledge on all aspects of integrating architectural design with technical and environmental building solutions towards the challenges of a better world. Provocative and inspirational, each volume in the Series aims to stimulate theoretical and creative advances and question the outcome of technical innovations as well as the far-reaching social, cultural, and environmental challenges that present themselves to architectural design today. Each brief asks why things are as they are, traces the latest trends and provides penetrating, insightful and in-depth views of current topics of architectural design. *Springer Briefs in Architectural Design and Technology* provides must-have, cutting-edge content that becomes an essential reference for academics, practitioners, and students of Architecture worldwide.

Shveta Mohan · Vijayalaxmi J.

Embodied and Operational Carbon in Buildings

Strategies to Decarbonize

Springer

Shveta Mohan
Sustainability, WSP Consultant
Bangalore, Karnataka, India

Vijayalaxmi J.
Department of Architecture
School of Planning and Architecture
Vijayawada, Andhra Pradesh, India

ISSN 2199-580X ISSN 2199-5818 (electronic)
SpringerBriefs in Architectural Design and Technology
ISBN 978-981-97-7186-8 ISBN 978-981-97-7187-5 (eBook)
https://doi.org/10.1007/978-981-97-7187-5

This Springer imprint is published by the registered company Springer Nature Singapore Pte Ltd.
The registered company address is: 152 Beach Road, #21-01/04 Gateway East, Singapore 189721, Singapore

Ms. Shveta Mohan dedicates the Book to her grandfathers.
Dr. Vijayalaxmi J. dedicates the Book to all her students.

Foreword

The book *Embodied and Operational Carbon in Buildings* by Ms. Shveta Mohan and Prof. Vijayalaxmi J., is a must-read book recommended to all building, energy, and environmental professionals, decision-makers, and students. It addresses and provides insights and useful scientific and engineering solutions for a very important problem in the built environment: 'Embodied and Operational Carbon and Emissions'. Considering that the energy efficiency of buildings is improving fast, and the operation energy consumption of the building sector is expected to decrease significantly soon, embodied energy and embodied carbon will be one of the principal terms in the energy and environmental balance of buildings. It is, therefore, necessary to understand the science and engineering related to the embodied and operational carbon, to develop methodologies to assess its magnitude, to evaluate its impact under various local and regional conditions, and most importantly to develop, implement, and assess appropriate solutions to minimize or at least reduce significantly its impact. The present book is among the very few documents that address the problem. It provides comprehensive and well-organized knowledge and information and succeeds in informing and educating the reader on almost all important aspects of the problem and finally offering a complete picture of the necessary scientific background including both the existing knowledge and future scientific developments.

M. Santamouris

Mat Santamouris
马特.桑塔莫里斯教授
Scientia Professor
Anita Lawrence Chair High Performance Architecture
School Built Environment
Faculty Art Design and Architecture
University New South Wales
Sydney, Australia

Preface

The Emissions Gap Report 2023 states that temperature records have escalated and climate impacts are to intensify as greenhouse gas emissions hit new high. The SDG 13 seeks to take urgent action to combat climate change and its impacts. Concentrations of greenhouse gases continued to increase in 2020, reaching new record highs. It was one of the three warmest years on record, with the global average temperature about 1.2 °C above the 1850–1900 baseline. It is imperative for all countries to do their part to meet the commitments to reduce carbon emissions. India is committed to reducing CO_2 emissions by 1 billion tons by 2030; reducing carbon intensity below 45% by 2030; and finally achieve a Net-Zero emission target by 2070. In the building sector, energy consumption and carbon emissions in the built environment are largely born out of the use of electricity and building materials. In India, under the current scenario, embodied and operational emissions have a share of 40% and 60%, respectively. India's total building floor area is predicted to be around 57.6 billion m^2 by 2050 from 15.8 billion in 2015. With the conventional building systems, this will escalate the demand for construction materials like cement, steel, bricks, glass- and all of these are energy and emissions intensive materials. Hence, there is a need to reduce the use of carbon intensive materials and adopt low carbon emitting materials.

This book offers the basics of embodied and operational carbon while discussing the inclusion of carbon emission in the GBRS at global and national level. This book also critically explores the important topic of embodied and operational carbon of buildings with insights on the strategies to measure and reduce embodied carbon in buildings through a case study and application approach. This approach assesses the impact of embodied carbon on the choice of structural systems, alternate building materials, alternate building technologies, and air conditioning system. The impact of these alternate measures in reducing embodied and operational carbon is analyzed by demonstrating its use on a base case building.

This book is written with a hope that architects, construction engineers, interior designers, architecture students, civil engineering students and all those associated with the construction industry gain an insight into their roles in reducing carbon emissions in order to achieve the National and International commitments.

Dr. Vijayalaxmi J.
Professor
Department of Architecture
School of Planning and Architecture
Vijayawada, Andhra Pradesh, India

Contents

Abbreviations

AAC	Autoclaved Aerated Concrete
ASHRAE	American Society of Heating, Refrigerating, and Air Conditioning Engineers
BIM	Building Information Modelling
BLA	Bamboo Leaf Ash
BREEAM	Building Research Establishment Environment Assessment Methodology
BUA	Built Up Area
CBAM	Carbon Border Adjustment Mechanism
CF	Carbon Footprint
CFC	Chloro-Fluoro Carbons
COP	Coefficient of Performance
CoP	Conference of Parties
CSEB	Compressed Stabilized Earth Blocks
DGU	Double Glazed Unit
EAC	Energy Attribute Certificate
EC	Embodied Carbon
EC3	Embodied Carbon Calculator
EC3	Embodied Carbon in Construction Calculator
ECBC	Energy Conservation Building Code
EE	Embodied Energy
EOL	End of Life
EPD	Environmental Product Declaration
EPI	Energy Performance Index
EPI	Energy Power Intensity
EUI	Energy Use Index
FaL-G	Fly Ash Lime Gypsum Blocks
FSC	Forest Stewardship Council
GBRS	Green Building Rating Systems
GGBS	Grund Granulated Blast-furnace Slag
GHG	Greenhouse Gas

GRIHA	Green Rating for Integrated Habitat Assessment
GWP	Global Warming Potential
HVAC	Heating, Ventilation and Air Conditioning
IEQ	Indoor Environmental Quality
IGBC	Indian Green Building Council
ILFI	International Living Future Institute
LBC	Living Building Challenge
LCA	Life Cycle Assessment
LED	Light Emitting Diode
LEED	Leadership in Energy and Environmental Design
LETI	Low Energy Transformation Initiative
LPD	Lighting Power Density
OC	Operational Carbon
OE	Operational Energy
OPC	Ordinary Portland Cement
OT	Operative Temperature
PFA	Pulverized Fuel Ash
RCC	Re-in forced Cement Concrete
RE	Renewable Energy
RF	Radiative Forcing
RHA	Rice Husk Ash
SHGC	Solar Heat Gain Coefficient
SRI	Solar Refractive Index
TERI	The Energy and Resources Institute
UHIE	Urban Heat Island Effect
USGBC	U.S. Green Building Council
VAV	Variable Air Volume
VLT	Visible Light Transmission
VRF	Variable Refrigerant Flow
WLC	Whole Life Carbon
WWR	Window Wall Ratio
ZNC	Zero Carbon Certification

List of Figures

List of Tables

Chapter 1
Embodied and Operational Carbon Emission

1.1 Introduction

The Paris Agreement treaty which was adopted at the Conference of Parties (COP21) states that global temperature rise must be limited to *1.5* °C by the end of the century to avoid the drastic, critical impacts and risks to our environment [1]. The primary reason for this temperature rise is the emission of greenhouse gases (GHG) which trap heat in the atmosphere and increase the temperature. Out of these GHGs, CO_2 has the highest positive radiative forcing (RF) and is more abundant in the atmosphere. It also remains in the atmosphere longer than other major heat-trapping gases and is predominantly generated by human activities.

Since the Paris Agreement, there have been annual conferences of parties (COPs) to track, discuss, and target actions to reduce carbon emissions by each country [2].

However, despite this, the IPCC concluded that from the implementation of the current Nationally Determined Contribution (NDCs), it may not be possible to limit global warming to 1.5° with limited or no overshoot. Maintaining the possibility of achieving the Paris agreement goal depends crucially on strengthening ambition and implementation to reduce carbon emissions this decade [3].

This is further reflected in the Buildings and Construction sector. The overall energy demand and emissions of the building sector have continued to grow, rising by 1% from 2021. It contributes to about 21% of global GHG emissions [4]. COP28 saw the launch of the Buildings Breakthrough and the 'Buildings Breakthrough Agenda Report', and this includes 28 countries, where operational buildings emissions must decrease by approximately 50 percent from their 2022 levels by 2030.

Actions at the national level that support the decarbonization of the existing and new buildings sector are critical to the delivery of this objective and the Paris agreement [4].

It is crucial for countries to redouble efforts to integrate best practices of improved building performance, low-carbon construction, and undertaking efforts to improve the resilience and climate adaptation of existing and new buildings. It is particularly

S. Mohan and Vijayalaxmi J., *Embodied and Operational Carbon in Buildings*,
SpringerBriefs in Architectural Design and Technology,
https://doi.org/10.1007/978-981-97-7187-5_1

relevant as the building sector will only expand further, with more focus on emerging and developing economies. Therefore, taking early action to prevent an increase in emissions is imperative.

1.2 Types of Emission

The terminology for the definition of 'energy' and 'carbon' which are referred to throughout this book, are as follows:

Energy is the capacity to do work across the life cycle of the building or product. This includes Production, Transport, Manufacture, Construction, Use, Demolition and Disposal. Energy can be fossil fuel based or from renewable sources.

Carbon is 'carbon dioxide' or 'equivalent carbon dioxide' emitted during a building's life cycle. The carbon emissions can be related to 'energy' particularly when fossil fuel sources are used. Energy derived from a high fossil fuel mix will lead to high-carbon emissions and renewable energy production will have low-carbon emissions. This book is particularly focused on carbon emissions.

The broad categorization of 'carbon emissions' is based on the life cycle stages of a building into two major types.

Figure 1.1 shows the life cycle of the building with the following modules.

1. Product (A1-A3) includes extraction and upstream processing of materials, transportation, and manufacturing impacts.
2. Construction (A4-A5) includes transport to the site.
3. Use (B) includes maintenance and replacement.

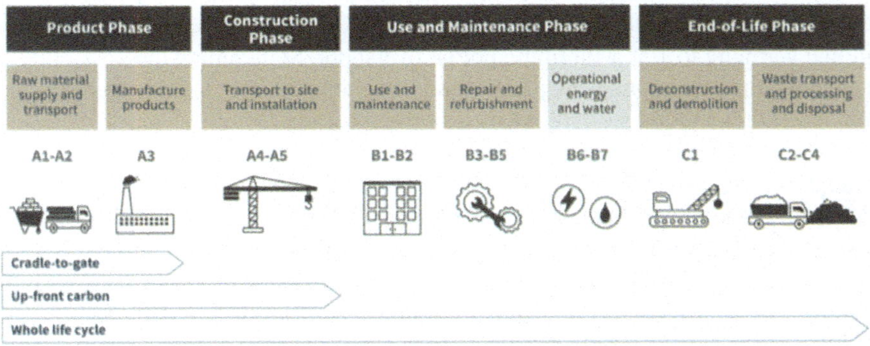

Fig. 1.1 Carbon emissions boundary based on the life of the building from RMI

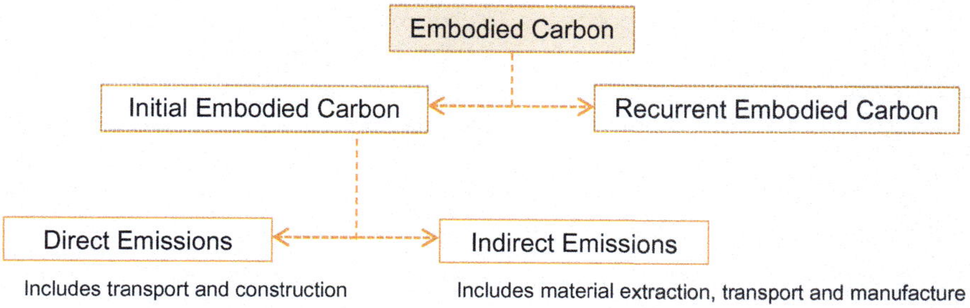

Fig. 1.2 Subdivision of initial or upfront embodied carbon emissions

4. End of life (C) includes demolition and disassembly as well as waste processing and transportation.

- *Embodied carbon emissions (EC)*

Embodied carbon is the sum of greenhouse gas emissions released during raw material extraction, transportation, manufacturing, construction, maintenance, renovation, and end-of-life for a product or system comprising A1-A5 in Fig. 1.1.

Within the life cycle modules, embodied carbon emissions can be further organized into the following phases.

– *Upfront carbon*: The emissions caused in the material's production and construction phases (A1-A5) of the lifecycle before the building or infrastructure begins to be used are upfront carbon. In contrast to other categories of emissions listed here, these emissions have already been released into the atmosphere before the building is occupied or the infrastructure begins operation.

This can be divided further into direct emissions and indirect emissions as seen in Fig. 1.2.

– *Cradle-to-gate:* The emissions emitted in the extraction, and manufacture of the raw material required for the building are included in the life cycle boundary for cradle-to-gate emissions.
– *End of Life: The carbon emissions associated with deconstruction/demolition, transportation from the site, waste processing, and disposal routes (C1-C4) of a building or infrastructure lifecycle, which occurs after its use is End-of-life emission.*
– *Whole life embodied carbon: The embodied carbon across all modules of A1-C4 excluding B6-B7 are considered under whole life embodied carbon of buildings as seen in* Fig. 1.1.
– *Beyond life cycle:* Module D refers to the end of life of the material, and factors in recycling or reuse as well as acknowledging waste as a resource, which is crucial for embodied carbon reductions. However, this is not currently mandatory for reporting embodied carbon emissions.

This carbon emission is reported as Global Warming Potential (GWP) and is measured *relative* to the impact of one molecule of carbon dioxide, usually over a 100-year timeframe. One kilogram of carbon dioxide has a GWP of 1 $kgCO_2e$, whereas, for example, one kilogram of methane is approximately 28 $kgCO_2e$, where 'e' is the equivalent impact indicator.

This allows a common unit of measure for all greenhouse gases (GHG) for comparability, impact studies, and target setting.

- *Operational carbon (OC)*

These include all carbon emissions during the use phase of the building, which includes heating, hot-water production, cooling, lighting systems, equipment, and lifts. It is indicated as B6-B7 in Fig. 1.1.

1.2.1 Key Difference

The key difference between embodied and operational emissions is that, while operational outputs are a larger portion of the emissions, they can be reduced across the lifespan of the building by using effective energy reduction strategies, measures, and offsets. Embodied emission outputs, however, need to be strongly addressed in the design and construction stage, to optimize the real energy and carbon savings over a period of time [5].

The World Green Building Council has stated its vision of achieving net-zero by 2050, supported by a 40% embodied carbon reduction. It also highlights that despite the ambitious targets, embodied energy and carbon have not been given the same importance as operational energy and carbon in terms of reduction and mitigation measures [6].

This is also supported by the number of publications as there are 17,100 publications on embodied carbon compared to the 21,900 research articles on operational energy [7].

Therefore, to address the growing concerns of embodied carbon reductions, the methodology for assessing and monitoring embodied carbon emissions should be established.

1.3 How is Embodied and Operational Carbon Quantified?

Life cycle assessment (LCA) is the most refined and comprehensive methodology used for assessing environmental impacts associated with buildings at present [5].

As depicted in Fig. 1.1, the analysis of the environmental impact of a product, process, or measurable activity is defined as life cycle analysis [8].

1.3.1 Metrics Used to Quantify Carbon Emissions

The common metric used to quantify carbon emissions is global warming potential (GWP). It allows for the comparison of gases by measuring the amount of energy one ton of a gas absorbs over a 100-year period in relation to one ton of carbon dioxide (CO_2). A higher GWP indicates a more pronounced warming effect compared to CO_2. This standardized metric assists in evaluating the diverse contributions of various greenhouse gases to the phenomenon of global warming.

In the case of operational carbon emissions, the same metric of GWP is applied and can be quantified over various parameters like area or time, for example, $kgCO_2e/m^2/year$, whereas embodied carbon is predominantly quantified over an area of the building assessed as $kgCO_2e/m^2$.

1.4 Methodology of Carbon Emissions Calculations

The calculation methodology for carbon emission follows the simple process of multiplication using volume or quantity and carbon emission factors. This is shown in Fig. 1.3.

This calculation can be employed in various ways. The predominant methods are:

1. Numerical method.
2. Software-based method.

The variable with the largest impact on embodied carbon is the carbon emission factor of the building materials. There are also other variables and parameters that similarly impact embodied and operational carbon emissions and are discussed in Sect. 1.4.1.

$$\text{Total embodied carbon of a building per year } (kgCO_2e/m^2) = \frac{\text{Mass of materials (kg)} \times \text{Carbon emission factor } (kgCO_2e/kg)}{\text{Total built up area of building}}$$

$$\text{Total operational carbon of a building per year } (kgCO_2e/m^2/year) = \text{Energy use intensity } (KWh/m^2/year) \times \text{Carbon emission factor } (kgCO_2e/KWh)$$

Fig. 1.3 Simplified equation to calculate embodied carbon

1.4.1 *Variables Which Affect Embodied and Operational Carbon Emissions*

Key decisions regarding these variables and associated boundaries will have an impact on embodied and operational carbon emissions as discussed in Table 1.1 [9].

Majority of the variables described in Table 1.1 are decision-based variables, implying that they can have a great impact before the construction, or operations are in place. It is important to build this carbon-based consciousness right from the design stage and through the life of the building to ensure a significant impact in carbon reductions measured.

1.4.2 *Possible Mitigation Measures for Carbon Emissions*

The possible carbon reduction measures should address the variables in Table 1.1 to optimize the various spatial, temporal, and physical characteristics of the study. This can be achieved by using the hierarchy defined in Fig. 1.4.

- **Build nothing**:
  These mitigation strategies address the initial boundary and necessity for development. Where possible, the existing building can be reused or repurposed. This leads to less new material and a reduction of the associated emissions.
- **Build less**:
  These strategies address reduction in requirements to only meet the necessity and not more. This can be exemplified by removing unnecessary bedrooms or limiting the built-up space, etc.
- **Build clever**:
  Smart building strategies reuse and are adaptable to various scenarios to reduce carbon emissions. For example, reuse of wood shutters and doors, can after the life of the building, be sent to the scrap yard or salvaged. This could also be modified to suit the different uses in new buildings such as commercial, residential, etc.
- **Build efficiently**:
  Efficient build strategies involve a reduction of carbon loads through design and materials. This is an effective measure for employment as a design guideline as they target carbon-conscious design at early stages.
- **Waste Reduction**:
  Waste reduction activities across the life cycle from manufacture to use and disposal can effectively reduce carbon emissions. For example, to reduce construction waste by reuse of rubble can form effective mitigation strategies.

As seen in Fig. 1.4, majority of the carbon reduction strategies can be employed before the construction of the building. This implies the early design stage decisions for a proposed development are crucial to the carbon reduction strategy.

Table 1.1 The variables which affect carbon emission are summarized below

	Category	Variable	Summary
1	Temporal differences	Life cycle stages	They are dependent on the boundary of the study set in the Life Cycle Assessment. For example, upfront embodied carbon, etc.
		Service life of the building	The life of the building particularly affects the operational carbon emissions significantly
2	Space of building considered	Building elements	The building elements considered such as shell and core, finishes, etc. significantly impact embodied and operational carbon
3	Physical characteristics	Building structure	The structure of the building namely concrete or steel, timber, masonry, etc.
		Building materials	Building materials such as brick, glass, steel, etc., and their properties such as density, etc. also have an impact on carbon emissions
		Building typology	Residential, commercial, etc. will have different design approaches and different targets for emissions
		Building location	The building location affects design based on climate which can have an impact on both embodied and operational carbon
		Building density	The density of the building, low-rise, high rise can define the structure and design, which also affects carbon emissions
4	Procedure definitions	Modeling approach	The type of approach to LCA like process-based, etc. plays a role in the impact of emissions
		Width of input	The boundary definitions such as processes involved, products and activities, etc.

(continued)

Table 1.1 (continued)

	Category	Variable	Summary
		Data sources for emission factors	Government sources or international/global data, literature based or any other credible sources
		Functional unit	The area of the building or the unit for the LCA study such as GIA, BUA, 1 component, whole building, etc.

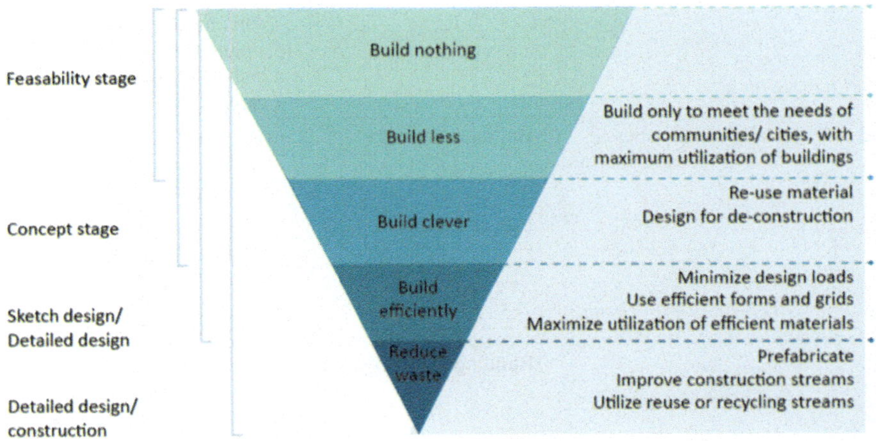

Fig. 1.4 Figure describing pathway to reduce embodied carbon

1.5 Early Design Stage

According to [10], early material-related emissions form 15% of a building's total emissions at the concept stage. Almost 50% of embodied emissions come from structural systems at a concept and developed design phase. Another 15% of embodied emissions come from material selection. [10].

In most cases, the utilization of LCA is a highly intensive data-driven process. Though the early design stage has a substantial carbon impact, it does not have precise building data, since the design variables are not completely set [11].

Out of the design decisions taken in an early stage with respect to efficient builds, past studies have found that structure and materials have the most impact on the embodied carbon emissions of a building [12, 13].

A few case studies of the structural impact of buildings are summarized in Table 1.2

Table 1.2 Case studies of structural impact on buildings

S.No	Typology	Source	Location	LCA boundary	Height	Structural material	Percentage of EC in building (%)
1	Residence	[13]	Cuntan, Chongqing, China	Cradle to gate	28 floors	RCC and shear wall structure	78.41
2	Office	[14]	China	Cradle to gate	13 floors	Steel and concrete	65.75
3	Residence	[15]	Kerala, India	Cradle to gate	4–11 stories group	RCC structure	76.65
4	Residential buildings	[16]	China	Cradle to gate	–	RCC structure	80
5	Office buildings	[17]	USA	Cradle to gate	5 stories	RCC structure	71

The most efficient low embodied carbon strategies include reducing area and material use, applying reused and recycled materials, using materials with very low-carbon emission, sourcing local materials, and adopting materials with high durability and long service life [18].

All these decisions can be taken at a decision-making design stage to optimize embodied carbon reduction.

However, it should be noted that carbon calculation and emissions vary geographically due to the carbon emission factors and other dependent variables listed in Table 1.1.

1.6 Summary

To align with the Paris Agreement goals, buildings and construction sector emissions must be reduced by an average of 9% per year till 2030. This will result in a more than 50% reduction by the end of the decade. Alongside building operations emissions, embodied carbon emissions need to be reduced through improved manufacturing processes and the use of renewable energy in their production [4].

Due to rapid urbanization, construction will be on the rise, and it is imperative that embodied and operational carbon needs to be equally targeted and reduced effectively to meet the Paris Agreement targets.

National efforts are crucial to reach these goals of net-zero carbon emissions. The implementation of stronger policies to enhance energy efficiency and address carbon emissions from building materials and construction is required.

In India, the buildings sector currently accounts for 40% of national CO_2 emissions, but 70% of expected Indian buildings have yet to be constructed [19]. It is most important, for such a developing country to take early action in the context of carbon emissions from buildings to aid sustainable growth.

Due to its large economy, India has been one of the world's largest emitters of greenhouse gases in recent decades, although per capita emissions are below the global average. In 2020, India's top greenhouse gas emitting sectors are power generation, iron and steel manufacturing, and road transport. Following the COVID-19 pandemic, the carbon share of power generation in India has reached a record high in 2021, pushing emissions 13% above 2020 levels. Coal remains a major contributor to India's greenhouse gas emissions. Eliminating carbon from power generation and industry is a major challenge for India to align its greenhouse gas emissions with its current growth ambitions. India's Nationally Determined Contributions (NDC) targets were set in 2015 to achieve by 2030, with an aim to.

i. reduce emissions by 33–35% below 2005 levels.
ii. achieve 40% non-renewable energy in electricity generation.
iii. creating a carbon sink of 2.5–3 GtCO2e through forest cover.

At the 26th meeting of the United Nations Framework Convention on Climate Change (COP 26) in November 2021, India announced its goal of becoming net neutrality by 2070. India said it would reduce its gross domestic product (GDP) output by 33–35% by 2030. India has also pledged to generate 50% of its electricity from non-fossil fuel sources by 2030 and re-affirmed its commitment to protecting forest cover, which acts as a carbon sink. It also aims to generate 500 GW of energy from non-fossil fuel sources by 2030. This can be done through the implementation of policies and guidelines, conscious design and implementation of new buildings, use of low-carbon materials, and decarbonization of construction processes. This book becomes significant in the light of implementing India's long-term low-carbon development strategy.

References

1. Masson-Delmotte V, P Z, HO, P, D, R, J, S, PR, S, T, W (2018). Global warming of 1.5°C.An IPCC special report on the impacts of global warming of 1.5°C above pre-industrial levels and. IPCC
2. UNFCCC (2018) Key aspects of the paris agreement. Retrieved from United Nations Climate Change: https://unfccc.int/most-requested/key-aspects-of-the-paris-agreement#:~:text=The%20Paris%20Agreement's%20central%20aim,further%20to%201.5%20degrees%20Celsius.
3. Olhoff A, Christensen J (2023) Emissions gap report 2023: broken record. Nairobi: UNEP
4. United Nations Environment Programme (2024) Global status report for buildings and construction. Nairobi: UNEP
5. Pomponi F, Moncaster A (2018) Scrutinising embodied carbon in buildings: The next performance gap made manifest. Renew Sustain Energy Rev, 2431–2442
6. World Green Building Council (2019) Bringing embodied carbon Upfront. Ramboll and World GBC
7. Google scholar (2023) Google scholar. Retrieved from https://scholar.google.co.in/scholar?hl=en&as_sdt=0%2C5&as_ylo=2023&q=energy+building&btnG=

8. Golisano Institute for Sustainability (2020) What is life cycle assessment (LCA)? Retrieved from RIT educational blogs: https://www.rit.edu/sustainabilityinstitute/blog/what-life-cycle-assessment-lca#:~:text=An%20LCA%20is%20a%20systematic,that%20make%20up%20industrial%20production
9. Pan W, Teng Y (2021) A systematic investigation into the methodological variables of embodied carbon assessment of buildings. Renew Sustain Energy Rev, 141
10. Hakkinen T, Kuittinen M, Ruuska A, Jung N (2015) Reducing embodied carbon during the design process of buildings. J Build Eng, 1–13
11. Marsh R, Rasmussen NF, Birgisdottir H (2018) Embodied carbon tools for architects and clients early in the design process. In Pomponi F, De Wolf C, Moncaster A. Embodied Carbon Build-Mitig, Meas Manag (pp 167–190). Springer
12. Gibbons O, Orr J (2022) How to calculate embodied carbon. Inst Struct Eng
13. Li S, Yan H, Chen J, Shen L (2017) A life cycle analysis approach for embodied carbon for a residential building. In: Proceedings of the 20th International Symposium on Advancement of Construction Management and Real Estate (pp 1185–1196). Researchgate
14. Luo Z, Yang L (2015) Embodied carbon emissions of office building: a case study of China's 78 office buildings. Build Environ
15. Kurian R, Kulkarni KS, Ramani PV, Meena CS, Kumar A, Cozzolino R (2021) Estimation of carbon footprint of residential building in warm Humid Climate of India through BIM. Energies, 14
16. Cang Y, Yang L, Luo Z, Zhang N (2020) Prediction of embodied carbon emissions from residential buildings with different structural forms. Sustain Cities Soc
17. Junnila S, Horvath A (2003) Life-Cycle environmental effects of an office building. J Infrastruct Syst, 4
18. Kjendseth M, Fufo SM, Andresen I (2018) Design strategies for low embodied carbon in building materials. In Pomponi F, De Wolf C, Moncaster A. Embodied Carbon Build—Meas, Manag Mitig (pp 323–339). Springer
19. GBPN (2022) Towards zero carbon: building policies In India 2022. GBPN

Chapter 2
Embodied Carbon Impacts of Green Building Rating Systems

2.1 Introduction

India is the fourth fastest-growing economy globally and has seen a combined growth in terms of rapid urbanization and population in the last few decades.

In India, the CO_2 emission has tripled since 1990. In order to meet the targets of the Paris Agreement treaty, which was re-affirmed at the Conference of Parties (COP26) held in Glasgow, India has announced the net-zero emissions target by 2070 for the country. India will also achieve a carbon intensity reduction of 45% over its 2005 levels by 2030 while meeting 50% of its energy demand through renewables [1].

However, in India, as a consequence of rapid urbanization, the current population in urban areas is predicted to increase by 65% in 2050 [2]. This will lead to a considerable and sudden burst in urban development. As discussed, 70% of the buildings associated with this predicted development are yet to be built [3]. Therefore, there is a huge potential in the field of architecture, planning, and energy efficiency to look at effective strategies and mitigation measures to reduce the carbon emissions of the building industry in order to meet the required national emissions targets.

2.2 Indian Scenario

India is the world's third largest emitter of carbon, after China and USA. In COP26, India has pledged to aim for net zero emissions by 2070. A major path to achieving this goal would be a reduction in embodied energy of materials and operational energy [4].

The regulations in India with respect to decarbonization initiatives are primarily focused on addressing operational carbon, with less initiatives on the life cycle approach for embodied carbon. Standard codes like the Energy Conservation

S. Mohan and Vijayalaxmi J., *Embodied and Operational Carbon in Buildings*,
SpringerBriefs in Architectural Design and Technology,
https://doi.org/10.1007/978-981-97-7187-5_2

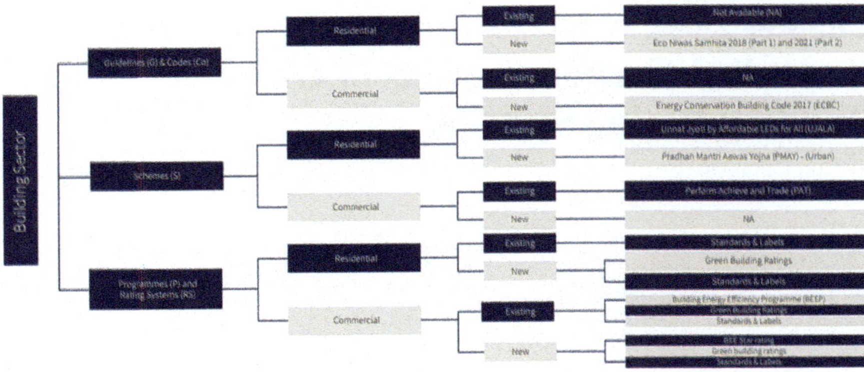

Fig. 2.1 Guidelines, codes, programs, schemes, and rating systems for green/energy efficient building sector [6]

Building Code (ECBC) and Eco-Niwas Samhita (ENS) measure building performance only based on the operational energy use of the buildings [5]. Apart from these, the government has developed guidelines, codes, schemes, and programs for building decarbonization as indicated in Fig. 2.1.

Annual energy consumption of construction materials in India is exceeding 2 billion tons. Further, the energy expended for the manufacture of building materials constitutes 20–25% of India's total energy demand [7].

In the Indian context, the carbon emission factor of purchased electricity is 1 KWh = 0.82 kgCO2e/KWh, where the factor of 0.82 accounts for the fuel mix for electricity generation in India, which comprises 68% thermal power, 17% hydropower, 12% renewable power and 2% nuclear power and 27% transmission losses [8].

Research studies performed in different countries differ from one another in terms of data relating to raw material quality, production processes, economy, delivered energy generation, transportation distances, energy use in transport, and human labor [9].

In terms of carbon emission factor databases, The International Finance Corporation (IFC) has also created an Indian Construction material database of Embodied Energy and Global Warming Potential for the common materials used. These values have been aggregated using GaBi data, local knowledge, and stakeholder contribution [10].

The data marked as 'Rest of the World' or 'Global Averages' can be characterized to fit the Indian context. This can be used with BIM LCA which is combined with digital tools and model information to be aggregated to the Indian context.

Apart from the regulations discussed in Fig. 2.1, many developers and designers adhere to green building rating systems, which are voluntary and consistently updated. Therefore, the most fast paced and investigative integration of embodied and operational carbon reduction in buildings within the urban context of India is explored under green building rating systems.

2.3 Green Building Rating Systems

Green building rating systems (GBRS) are third party, voluntary, market-driven standards that measure the sustainability level of a building by using multiple predefined criteria. This encourages sustainable and environment-conscious design and could function as a driver for change in the Indian context, particularly relevant to carbon emissions.

India has two major green building rating systems namely the Indian Green Building Council (IGBC) and the Green Rating for Integrated Habitat Assessment (GRIHA), which rate buildings spread over India (Table 2.1).

Out of these, IGBC is the most popular rating system with over 2000 rated buildings [11].

2.3.1 IGBC

The Indian Green Building Council rating system itself is based on encouraging sustainable design and construction techniques in the development of new buildings. Points are awarded to projects based on performance related to sustainable site development, water savings, energy efficiency, material selection, and indoor environmental quality [12].

Table 2.1 Green building distribution in India based on location

	LEED-India Projects	% of total LEED-India Projects	GRIHA projects	% of total GRIHA projects	Total green projects	% of total green projects
Maharashtra	160	35.95	124	33.97	284	35.06
Tamil Nadu	80	17.97	12	3.28	92	11.35
Karnataka	49	11.01	18	4.93	67	8.27
Andhra Pradesh	42	9.43	17	4.65	59	7.28
Uttar Pradesh	23	5.16	23	6.3	46	5.67
Haryana	22	4.94	21	5.75	43	5.3
Delhi	7	1.57	34	9.31	41	5.06
Gujarat	16	3.59	20	5.47	36	4.44
West Bengal	9	2.02	13	3.56	22	2.71
Rajasthan	8	1.79	13	3.56	21	2.59

2.3.2 GRIHA

GRIHA is a quasi-public rating system for promoting sustainable building practices in India. It is voluntary and GRIHA specifically focuses on the sustainability of projects, more after they are constructed, as it requires a compilation of performance factors as part of GRIHA recognition. GRIHA utilizes a 1–5-star rating system that focuses on energy/power consumption, water consumption, water generation, and renewable energy integration. These factors are different than the LEED criteria and are tailored to suit the Indian context [13].

IGBC and GRIHA are the most often used rating systems in India. GRIHA is an easy but detailed rating system to achieve and requires time. IGBC uses per capita energy consumption, and this is an unreliable factor for rating, as India's per capita energy consumption is very low [12].

2.3.3 Case Study of the Embodied Carbon Impacts of Rating Systems

It has been stated in Sect. 2.2, that the Indian guidelines are not as focused on embodied impacts when compared to the operational carbon impacts.

For a better perspective, rating systems around the world are compared with the IGBC rating system from India, to review impacts from various criteria and its relevance to embodied carbon.

A research study by [14] established certain key criteria among international rating systems. These criteria are listed below.

1. Site.
2. Energy.
3. Water.
4. IEQ.
5. Material.
6. Waste and pollution.
7. Management.

A few rating systems across the world are chosen based on their geographic location and these criteria are then assessed. These rating systems include LEED from USA, BREEAM based in UK, Green Star in Australia and South Africa, Green Mark in Singapore, GBI from Malaysia, BEAM Plus from Hong Kong, IGBC from India, and CASEBEE from Japan.

The hierarchy of criteria importance is given in the following order.

The first to top is Energy followed by IEQ, which are dominant themes. Subsequently, water criteria are given importance. This is followed by materials and waste management which are given lower importance. This is the case for current rating systems which are wholistically targeted.

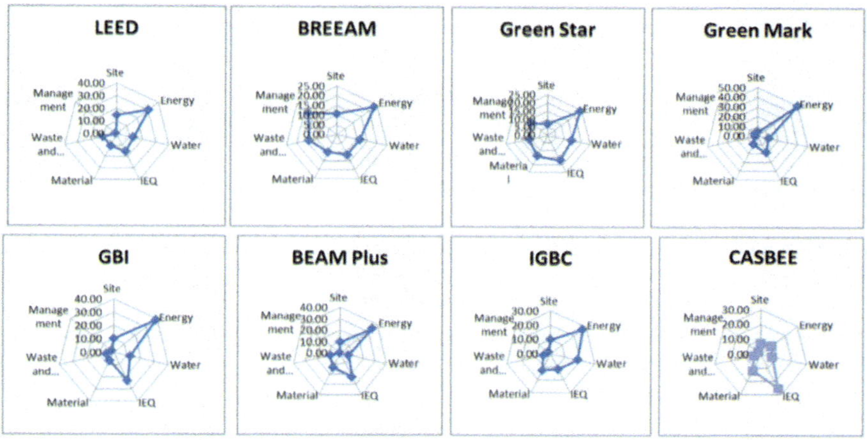

Fig. 2.2 Comparison of key credit criteria of rating systems [14]

The material criteria in the above GBRS are given a weightage of around 10% except GBI which is substantially lower at 6%. IGBC has a material weightage of slightly higher than 10%. So, it is not conclusive that Indian rating systems are poorer in their address of embodied carbon.

However, with increasing focus on carbon reduction strategies and rapid urbanization in India, 'materials' criteria will have a significant impact on the decarbonization of buildings. Future rating systems can focus on the weightage of these key criteria from Fig. 2.2 on the basis of the required carbon emission reduction for implementation.

2.3.4 Criteria for Study of Embodied Carbon in Rating Systems

Section 2.3.3 describes a study of the key criteria of rating systems, and it is found that materials and embodied carbon strategies overall have less weightage. For a more detailed methodology to study embodied carbon reductions in rating systems, the following method can be used.

One Click LCA categorizes these methods by increasing the order of efficiency into [15]

1. carbon reporting,
2. comparison in design,
3. carbon rating,
4. carbon caps, and
5. decarbonization.

Table 2.2 How each carbon reduction method functions in rating systems

Method	How it works	Examples
Carbon reporting	Calculate the project's embodied carbon and report it	EN15978, BREEAM Int'l
Carbon comparison	Compare design options for carbon, showing baseline and proposed with improvements against a self-declared baseline value	LEED v4, Greenstar, BREEAM UK
Carbon rating	Evaluation of carbon performance. Scale from best to worst, but no maximum value is applied. Fixed scale or clear methodology is absent	DGNB, BREEAM NL
Carbon cap	Calculate the project's embodied carbon and prove it is not exceeding the carbon limit	Energie Carbone, MPG
Decarbonization	Reduce carbon to a minimum, then compensate for all residual emissions by own energy export or buying offsets	Living Building Challenge, NollCO2

These parameters and how the reduction method works in rating systems are described in Table 2.2.

Out of these, carbon comparison particularly addresses the early stages of the project, while the remaining methods can be applied in various stages of a building's timeline.

Some of the recommendations include the following [15]

1. Target the early phase of the project.
2. Set an embodied carbon cap for common building types.
3. Apply a fixed method for setting the cap.
4. Incentives to achieve carbon reductions.
5. Set open compliance and make sure they are met.

These parameters are then applied to a series of relevant rating systems in India that address embodied carbon and materials to identify what strategies and methods have been used for carbon reductions in current green building rating systems.

2.4 Comparison of Rating Systems

Based on the criteria described in Sect. 2.3.4, the following six rating systems are reviewed to identify the extent of embodied carbon sensitivity (Table 2.3).

- IGBC (Indian Green Building Council).
- GRIHA (Green Rating for Integrated Habitat Assessment).
- LEED (Leadership in Energy and Environmental Design).
- EDGE (Excellence in Design for Greater Efficiencies).
- LBC (Living Building Challenge)—(International Living Future Institute).
- ZNC (Zero Net Carbon)—(International Living Future Institute).

Table 2.3 Comparison of rating systems from India—[12, 16]

Criteria studied	India	
	IGBC	GRIHA
Target the early phase of the project	It does not specifically address the early phase but includes passive architecture measures	Yes, using passive design
Setting the cap for Operational Carbon	Some strategies to address operational efficiency are: • >2% reduction from base case to be demonstrated using Passive Design • One or more passive strategies must be incorporated • High Solar Reflex Index (SRI) > 75% of exposed roof area (or) Vegetation > 50% (or) combination on roof • Ensure refrigerant, HVAC, and fittings are CFC free • Simulations at a comfort temperature of 24 ± 2 degrees (Adaptive Comfort) • Use ECBC for Window properties (SHGC), and opaque materials (U-value) • Lighting Power Density (LPD) is 20% over the ECBC baseline • Demonstrate Renewable Energy > 1% of total annual consumption on-site	Some strategies to address operational efficiency are: • Adoption of two to five passive strategies listed and reduction in UHIE • Peak heat gain—25 W/m^2 for Warm and Humid (Office Typology) • Energy Performance Index (EPI)—Benchmark at 90 kWh/m^2/year. Additional reduction up to > 50% (>45 kWh/ m^2/year) • Ensure HVAC, fire, insulation, refrigerant, CFC/ HFC free • Window wall ratio (WWR) < 60% and Visible light transmission (VLT)—0.27, SRR (Skylight Roof ratio) < 5% and SHGC < 0.35 • Shaded between 9 am March 15 to 3 pm September 15 • 60% adaptive comfort or implement seven natural wind (NV) passive strategies
Setting the cap for Embodied Carbon	• At least 2.5% of materials by cost are salvaged, reused, or refurbished • At least 10% of materials by cost are recycled content • At least 20% of sources are within 400 km • 50% of wood is rapidly renewable (or) FSC certified • At least 5 passive or active green materials (GreenPro)	• Replace cement 20% to > 30% with waste materials or 5% of aggregate with recycled aggregate • Select walls of nine options (low emission) • Replace 30% OPC within mortar • LCA benchmark reduction (unique base case) of superstructure 10% to > 15% • Other criteria deal with site and interiors • 8% of points for waste, recycled, and alternate materials
Compliance Criteria	• Passive Design—Optional • Energy Optimization—Partly mandatory • Materials criteria—Optional	• Passive Design—Optional • Energy Optimization—Partly mandatory • Materials criteria—Optional

(continued)

Table 2.3 (continued)

Criteria studied	India	
	IGBC	GRIHA
General range carbon values for offices with certification	• Embodied carbon—5–20% reduction in embodied carbon from conventional buildings • Operational carbon—40.5–113.4 $kgCO_2e$/sq.m/ year	• Embodied carbon—20% reduction in embodied carbon from benchmark buildings • Operational carbon—58.32–72.9 $kgCO_2e$/sq.m/ year
Strategies used in examples	A series of building examples are examined in Chapter 2 for IGBC and GRIHA as these are the primary rating systems used in the Indian context	

The tables above show Edge certification, LBC, and ZNC certification with high sensitivity to carbon strategies. However, LBC does not directly address carbon targets. The following differences were observed (Tables 2.4 and 2.5).

- In ZNC and Edge certification, embodied carbon is addressed separately from operating energy. Many of the certification systems have a materials section, which indicates some responsiveness toward embodied carbon, but addressing it separately with a target ensures the impact of the strategy.
- In Edge, it is mandatory to achieve a minimum value of embodied energy and carbon for certification. The same measures are implemented in the ILFI certification systems. However, low-carbon material strategies or LCA aren't a hard prerequisite in the other certification systems.

2.5 Summary

Green building rating systems are multi-criteria certifications for buildings, targeted at wholistic design and sustainability of the built environment. It varies primarily across geography, scale, and building type. These must have a comprehensive set of criteria to govern new construction, renovations, repairs, and demolitions. Extensive research on reducing operational energy consumption and the related greenhouse gas (GHG) emissions has correlated the importance of buildings in addressing climate change, leading to more energy-efficient codes. In this process, embodied carbon has been largely neglected from building rating systems. Even small embodied carbon reduction requirements can result in big reductions, because of the number of buildings that would be built in a developing country in the near future.

Currently, there is more focus on the operational phase of buildings. The operational emissions are likely to fall due to this emphasis along with the decarbonization of the electricity grid. On the contrary, embodied carbon is likely to enter an upward

Table 2.4 Comparison of rating systems from International—[17, 18, 21]

Criteria studied	International	
	LEED	EDGE
Target early phase of the project	Yes, with more focus on operational energy	Yes. Edge has three levels of certification. Level 1 at a preliminary stage, followed by Level 2 and Level 3 with more stringent carbon guidelines
Setting the cap for Operational Carbon	Some operational efficiency measures are: • Solar exposure, UHIE, temperature, rain, and winds are to be documented • Roof SRI 82 and 39 for high slope (or) green roof with native species • Comply with ASHRAE 90.1–2016 or USGBC guides for the Energy Performance Index (EPI) • HVAC refrigerant restriction with no CFC or HCFC • Performance cost index at least 5% to 50% improvement from the threshold • Lighting Power density (LPD)—15% to 45% reduction • Meet 70% of wall/ floor is mass wall/floor (or) 25% UA reduce (or) 50% SHGC reduction (with shading) • Points given based on percent Renewable Energy / Energy Attribute Certificates of total site used	Some operational efficiency measures are: • Level 1 EDGE Certified requires 20% or more savings in energy, water, and embodied energy in materials as a minimum requirement • Level 2 EDGE Advanced certification for buildings needs 40% or more on-site energy savings • Level 3 EDGE Zero carbon savings requires 100% renewables on-site or off-site or purchased carbon offsets to top off at 100%. All energy must be accounted for, including diesel and LPG
Setting the cap for Embodied Carbon	• GHG cost index at least 5% till 80% improvement from threshold • LCA for structure and enclosure from 5 to 20% reduction in more than two categories from baseline adhering to the guidelines of ISO 14044 • Use EPD for 20 products (or) two to five optimized EC products • At least 15–30% responsible producer, bio-based materials or FSC certified wood, other sustainable materials	• Edge certification clearly addresses embodied energy and emissions as it is required to achieve a percentage reduction of 20% for minimum certification In June 2023, it has also addressed embodied carbon caps in the materials category

(continued)

Table 2.4 (continued)

Criteria studied	International	
	LEED	EDGE
Compliance Criteria	• Passive Design—Optional • Energy Optimization—Partly mandatory • Materials criteria—Partly mandatory though LCA is not included under this	• Passive Design—Optional • Energy Optimization—Partly mandatory • Materials criteria—Partly mandatory All criteria are in compulsory percentage reductions
General range carbon values for offices	• Embodied carbon—56–250 kgCO2e/sq.m • Operational carbon— < 238 KWh/sqm/yr	• Embodied carbon—20% - reduction from conventional baseline • Operational carbon—not numerically specified
Strategies used in case examples	Harvard University, Boston LEED-certified buildings, where 22% of materials were manufactured within 500 miles and raw materials extracted in 500 miles	The Atria building in Bangalore uses efficient lighting and HVAC, along with high-performance glazing, which provides excellent insulation and materials which are low carbon and responsibly sourced

trajectory till 2040. The direction of green rating systems will be critical to the inclusion of embodied carbon, in the definition of a sustainable building. The USA has been proactive in including embodied carbon credits in the LEED Rating system since 2014. India too, needs to follow suit.

India is a developing country, therefore the construction industry emissions, which currently forms 1/6th of the country's GHG emissions, is expected to rise exponentially. This challenge needs to be tackled right from the national policy level and global organizations down to voluntary initiatives.

In terms of national regulations, India's National Action Plan on Climate Change (NAPCC) emphasizes sustainable development, addressing building aspects through the National Mission on Sustainable Habitat. India's National Determined Contribution (NDC) commits to reducing GDP emission intensity and increasing non-fossil energy capacity.

However, the building industry faces unique challenges like retrofitting existing infrastructure, initial costs, limited awareness, regulatory inconsistencies, and supply chain challenges. Future trends and changes involve the development of net-zero buildings, advanced materials, smart technologies, circular economy practices, and the strengthening of policies and incentives, which are largely tailored to the use phase of the buildings.

As more research inputs are shared with the building code developers, the pressure to include the latest in embodied energy reductions in the rating codes will increase as a response. Thus additional pathways for codes to reduce embodied carbon will become available. One of the most important ways to make the inclusion of Embodied energy reductions in building codes is to demonstrate evidence-based prescriptive strategies. This will instill confidence in the building industry stakeholders.

Table 2.5 Comparison of rating systems from ILFI—[19, 20]

Criteria studied	International	
	LBC—ILFI (No optional credits)	ZNC—ILFI
Target early phase of the project	Yes, must be net zero	Yes, must be zero carbon
Setting the cap for OC	• 70% reduction from an equivalent baseline building • All projects must account for total CO_2e in construction through sequestering materials or offsets • 105% of energy needs, through on-site renewable energy on an annual basis (needs to be metered)	• 100% of the operational energy of the project must be offset by on or off-site renewable energy • 25% reduction of energy use intensity from a new building that's ASHRAE compliant as baseline
Setting the cap for EC	• Show a 20% reduction in embodied carbon of primary materials from baseline (established using tools) • Labeled product used per 200 sq.m of area • 50% salvaged wood, rest should be low-risk materials • > 20% materials within 500 kms • No red list chemicals in 90% • One salvaged material per 500 sq.m of gross area	• 100% of the embodied carbon of materials and construction must be disclosed and offset • Embodied carbon emission of primary materials reduced by 10% from the baseline equivalent of size, and function • Total Embodied carbon (materials, transport, and construction [primary, interior]) < 500 $kgCO_2e/m^2$
Compliance Incentives	• Passive Design—Core Imperative • Energy Optimization—Core Imperative • Materials criteria—Core Imperative	• Passive Design—Core Imperative • Energy Optimization—Core Imperative • Materials criteria—Core Imperative
General range carbon values for offices	• Embodied carbon—At least 20% reduction from conventional baseline • Operational carbon—Net zero energy is mandatory	• Embodied carbon—0–500 $kgCO_2e$/sq.m • Operational carbon— < 178 kWh/sqm/year
Strategies used in case examples	Use of radiant cooling in Bullitt Center, Seattle helped in a significant EUI reduction of 12 Btu/ft2. It is 80–85% less than the average office performance in California It is built based on LBC by ILFI	London's Pancras Square shows a 16.8% reduction in embodied carbon from the baseline And 27% of components can be reused. Easy disassembly and flexibility

Therefore, green rating systems can be employed as a tool to address the gap in building carbon emissions, particularly concerning embodied carbon in India. The measurement tool identified and most often used is the Life Cycle Analysis (LCA). The green building rating systems are assessed along with LCA in the following ways: LCA integration, recognition and weightings of the building embodied and operational carbon emissions, and comparable metrics for the early design stage of buildings. In order to make evidence-based decisions on the use of building materials, building projects must report the building's GWP (CO2e). This can be done using a whole-building life cycle assessment (WBLCA) or whole life carbon assessment. The WBLCA can be used to compare a proposed design to a maximum CO2e limit per floor area or a GWP reduction as compared to a base case model. This can be similar to the energy use intensity (EUI) target used for operational energy comparison.

References

1. Ministry of External Affairs, GOI (2021). Media center. Retrieved from National Statement by Prime Minister Shri Narendra Modi at COP26 Summit in Glasgow: https://www.mea.gov.in/Speeches-Statements.htm?dtl/34466/National+Statement+by+Prime+Minister+Shri+Narendra+Modi+at+COP26+Summit+in+Glasgow
2. United Nations. (2014). World urbanization prospects, department of economic and social affairs. United Nations
3. Puri HS (2019) How India is building cities of the future. Econ Times
4. The Energy and Resources Institute (TERI) (2024). TERI's solutions for sustainable development. New Delhi: TERI.
5. Bhardwaj S, Tewari D (2022) Tackling embodied carbon from India's building sector. Retrieved from Alliance for an Energy Efficient Economy: https://aeee.in/tackling-embodied-carbon-from-indias-building-sector/
6. GBPN (2022) Towards Zero Carbon: Building Policies in India 2022. GBPN
7. Praseeda K, Reddy B, Mani M (2015) Embodied energy assessment of building materials in India using process and input–output analysis. Energy Build, 677–686
8. Central Electricity Authority, Ministry of Power (2022) CO_2 Baseline database for the Indian Power Sector. New delhi: Government of India, Ministry of Power
9. Lahiri A, Dixit S (2015) Alternatives to Cement in Concrete—A Review. Int J Sci & Eng Res
10. International Finance Corporation (IFC) (2017). India construction materials database of embodied energy and global warming potential. Eur Union IFC
11. Kanaujia V, Reddy AS, Kumar K (2017) Comparative review of Indian green building rating system. J Energy Res Environ Technol, 194–198
12. Indian Green Building Council (IGBC) (2016) IGBC green new buildings rating system. Hyderabad: IGBC
13. Smith MR (2015) "Green" building in India: a comparative and spatial analysis of the LEED-India and GRIHA rating systems. Asian Geogr, 73–84
14. Illankoon SI, Tam V, Le K, Shen L (2017) Key credit criteria among international green building rating tools. J Clean Prod, 209–220
15. One Click LCA (2018) The embodied carbon review. One Click LCA
16. GRIHA Council (2021) GRIHA V.2019—Third Edition. New Delhi: The Energy and Resources Institute (TERI)
17. LEED v4 (2021) Reference guide for building design and construction. Retrieved from USGBC: https://www.usgbc.org/guide/bdc#credit

18. UGreen (2023). EDGE Certification—The complete guide. Retrieved from UGreen: https://ugreen.io/edge-certification-the-complete-guide/#:~:text=Scope%3A%20EDGE%20Certification%20is%20focused,buildings%2C%20interiors%2C%20and%20neighborhoods
19. ILFI (2024) Living building challenge 4.1—Program manual. ILFI.
20. ILFI (2024) Zero carbon certification 1.1—Program manual. ILFI
21. EDGE Buildings (n.d.) Certification. Retrieved November 30, 2024, from https://edgebuildings.com/certify/certification/

Chapter 3
Life Cycle Assessment for Carbon Emissions

3.1 Life Cycle Assessment

Life cycle assessment (LCA) is a widely used method in the analysis of environmental impacts of products, technologies, or systems over the entire life cycle of a building. It helps to gain a thorough understanding of the material's impact on the environment and potential improvements at each step of the life cycle.

LCA is defined in ISO 14040 as 'A technique for assessing the environmental aspects and potential impacts associated with a product, by compiling an inventory of relevant inputs and outputs of a product system; evaluating the potential environmental impacts; and interpreting the results of the inventory analysis and impact assessment phases'.

LCA can be considered as a conceptual framework a tool, or even both, depending on the context of the study.

3.2 How and Why is LCA Important?

Businesses use LCA to display transparency and credibility to stakeholders and consumers in many ways.

- For existing products: As the environmental impact of products attains growing importance, businesses are starting to incorporate environmental aspects into the design and development process. This requires the identification of key environmental issues related to products at every stage of the life cycle.
- Research and development: When a product's environmental impact is critical to its future marketing or financially beneficial to an organization, life cycle assessment (LCA) is utilized in the research and development of new products.

S. Mohan and Vijayalaxmi J., *Embodied and Operational Carbon in Buildings*, SpringerBriefs in Architectural Design and Technology,
https://doi.org/10.1007/978-981-97-7187-5_3

- Other applications include strategic planning, public policy making, marketing, and others.

LCA impacts not only address carbon emissions through global warming potentials but can also address other environmental variables like ocean acidification, eutrophication, smog formation, etc. The carbon emissions are translated to global warming potential, through LCA by multiplying the masses with their characterization parameters.

However, it is to be noted that LCA addresses products only from an environmental perspective. Other aspects such as technical, economic, and social are not considered and should be integrated with the larger perspective for product design and development.

3.3 Standards for Life Cycle Assessment

The ISO 14000 is a family of Environmental Management standards that define how organizations and companies manage their environmental responsibilities.

Out of these,

- ISO 14040:2006: Life Cycle Assessment Framework: Defines the principles and framework of a Life Cycle Assessment.
- ISO 14044:2006 specifies requirements and provides guidelines for conducting an LCA study.
- Standards like ISO 14067 provide guidance to quantify the carbon footprint of products during an LCA.
- Additionally, standards like PAS 2050 and ISO 14020—ISO 14026 play a crucial role in preventing greenwashing and ensuring transparent communication of environmental footprint information.

Further ISO 14074 is being developed on guidelines, requirements, and principles for normalization, weighting, and interpretation of LCAs [1].

These standards define the phases and modules and offer a comprehensive view of the environmental impacts associated with all stages of a product's life, from raw material extraction to disposal. It forms the backbone of LCA studies.

3.3.1 LCA Modules

As discussed, and according to the ISO standards, the life cycle study of a building is divided into various phases based on the life of the building. These are defined as the modules of the Life Cycle Analysis study.

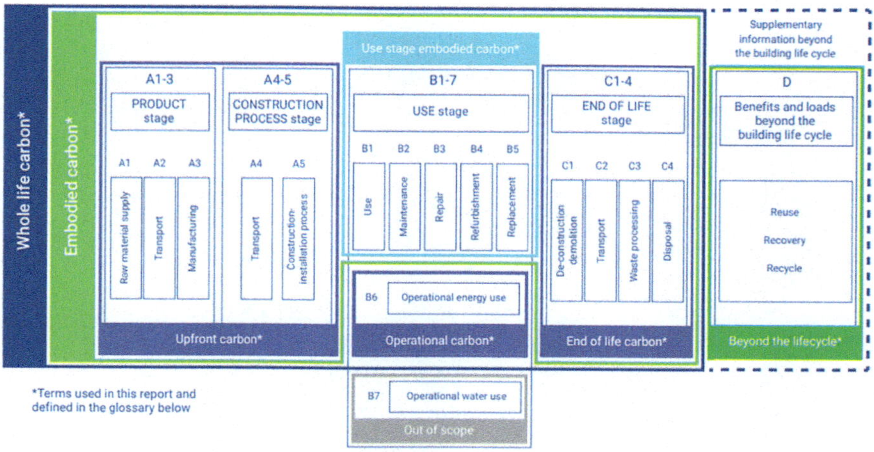

Fig. 3.1 Life cycle analysis modules [2]

The modules of LCA as categorized by BS EN 15978:2011 are:

1. Product or material phase [A].
2. Construction [A].
3. Operation [B].
4. End of Life [C].
5. Beyond the life cycle [D] as shown in Fig. 3.1.

The boundaries of the system are often crucial to the outcome of an LCA. These boundaries are defined by the stages described in Fig. 3.1. Stages A, B, and C form the whole life of a building and D is the module beyond the life cycle for EOL scenarios like recycling, waste, etc. The definition of scope is a crucial element of an LCA study and carbon emissions can be categorized across this.

The categorization of embodied carbon across the life cycle is illustrated in Fig. 3.2.

3.3.2 Calculation for LCA Study of Embodied Carbon

The life cycle assessment can be grouped into stages, based on the life cycle modules. This is demonstrated in Fig. 3.3.

- **Cradle-to-gate**—This scope covers A1 to A3 of extraction, transportation, and manufacture of the raw materials. It involves assessing a product until it leaves the factory gates, excluding the use and disposal phases. It is commonly used to assess embedded emissions in materials.
- **Cradle-to-site**—This scope covers A1 to A5 and goes beyond material to site-based transport and construction stage on the site.

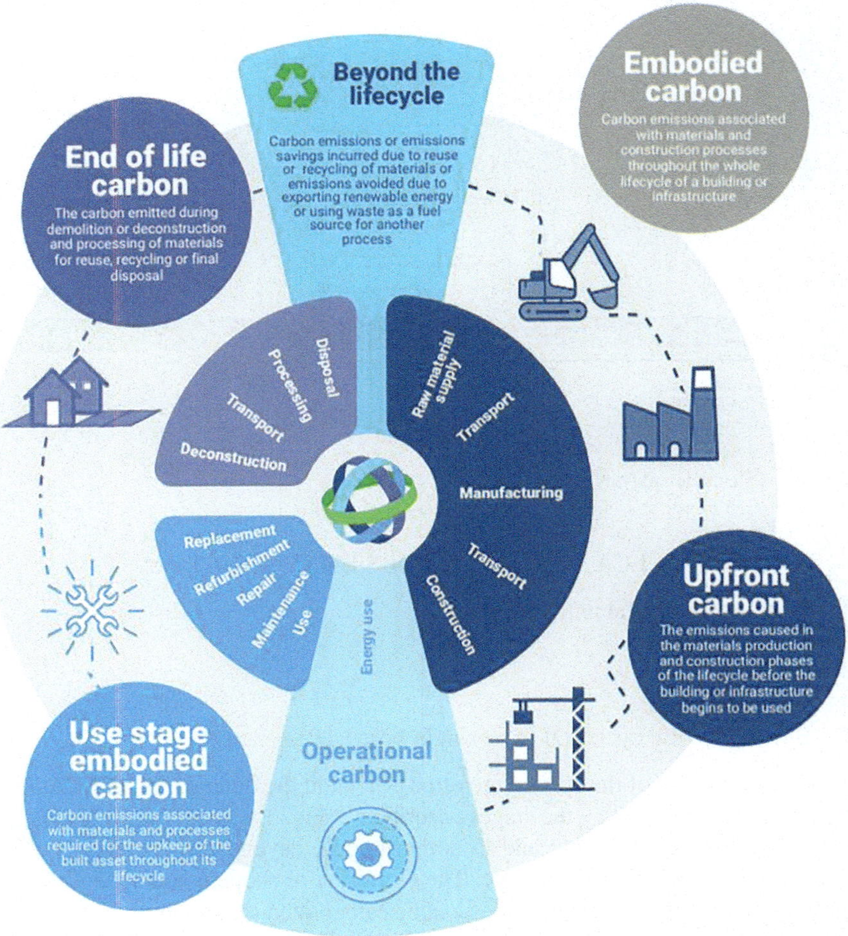

Fig. 3.2 Embodied emissions and definitions across the life cycle. Modules [2]

- **Cradle to grave**—This scope covers A1 to C4 and covers use phase and end-of-life considerations. This is important to create a holistic picture of the environmental impact of a particular subject [3].

The boundaries represented in Fig 3.3 include a cradle-to-cradle assessment which covers the complete depth of the life cycle including EOL scenarios. This is a concept within the Circular Economy, replacing the waste stage with a recycling process, creating a closed-loop system where materials are reused for new products. However, due to uncertainty of the treatment of the subject, availability of data, and other unknowns, the end-of-life calculations are often not considered.

Most often, Cradle-to-gate is the scope used for a quick analysis of embodied carbon, as it concerns approximately 65% of total embodied carbon emissions [4].

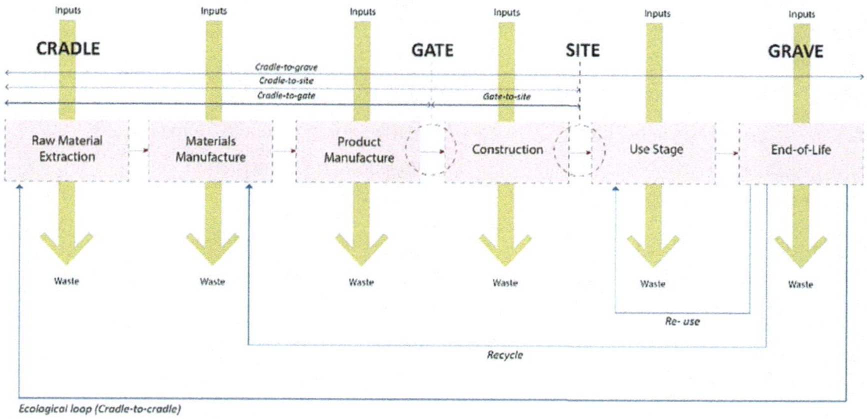

Fig. 3.3 Life cycle analysis modules and boundaries

3.4 What Are the Environmental Impact Indicators?

There are three primary areas of protection are taken into consideration while performing impact assessments in life cycle assessments. The three main areas of protection, which are as follows (Table 3.1):

1. Human health.
2. The natural environment (ecosystem quality), and
3. Issues related to natural resource use.

The five most often monitored environmental effect categories in LCA are:

- Global warming potential (GWP),
- Ozone depletion potential (ODP),
- Eutrophication potential (EP),
- Acidification potential (AP), and
- Smog formation potential (SFP).

The research behind generating these five impact categories has been established and reasonably consistent, which is why they are widely utilized in LCAs. Other impact metrics, such as human health effects and ecotoxicity, are less commonly employed because they were created with varied degrees of uncertainty.

3.5 Benefits of Using LCA

Life cycle assessment (LCA) offers certain advantages that can support sustainable development. A few of these advantages are as follows:

Table 3.1 Environment impact indicators related to LCA [5]

Environmental impact indicators		
Sl. no.	Areas of protection	Impact indicators
1	Human health	Global warming potential, human health
		Stratospheric ozone depletion
		Ionizing radiation
		Ozone formation, human health
		Fine particulate matter
		Human toxicity, cancer
		Human toxicity, non-cancer
		Water consumption, human health
2	Natural resources	Mineral resource scarcity
		Fossil resource scarcity
3	Ecosystem quality	Global warming, terrestrial ecosystems
		Global warming, freshwater ecosystems
		Ozone formation, terrestrial ecosystems
		Terrestrial acidification
		Freshwater eutrophication
		Marine eutrophication
		Terrestrial ecotoxicity
		Freshwater ecotoxicity
		Marine ecotoxicity
		Land use
		Water consumption, terrestrial ecosystems
		Water consumption, aquatic ecosystems

(1) Environmental effect reduction—Identification of environmental hotspots in a product's lifecycle, can help execute potential changes. This may help lower the environmental impact of the product.
(2) Resource preservation—Finding areas of inefficient resources and using more sustainable materials or methods as well as creating a sustainable supply chain can be done using LCA.
(3) Cost savings—Identifying and implementing more efficient processes and putting LCA recommendations into practice can save costs.
(4) Risk management—Any risks associated with a product's production, usage, and disposal can help advise risk management and action.
(5) Better relationships with stakeholders.

3.6 Challenges of LCA

(a) LCA may be economically unaffordable for the majority of organizations interested in using it due to the associated expenditures. This is a result of the time- and data-intensive nature of a complete life cycle assessment. An LCA will be more costly and time consuming the more thorough it is. The necessity for specialist knowledge and professional counsel during the impact and improvement analysis stages contributes to high expenses.
(b) Some life cycle assessments (LCAs) limit the system under consideration to a few key stages rather than the complete life cycle of the product or process under investigation.
(c) There is no singular or consistent methodology to conduct an LCA. Organizations have flexibility in perspective and implementation.
(d) Data from inventories can vary in quality.
(e) Certain research assumptions (such as defining the study's boundaries, selecting the data source, and determining the effect assessment method) may be arbitrary.
(f) The outcomes of this research may not be appropriate for local applications if they are centered on the national and regional levels.
(g) The availability and quality of pertinent data determine an LCA study's accuracy. The study's accuracy is constrained if the data are not sufficiently accurate.
(h) Furthermore, communicating the findings of an LCA study is difficult. In addition to differing research methodologies, life cycle assessment (LCA) generally yields outcomes about several environmental aspects. Product A might not always outperform Product B, but only in specific situations. Then, a more thorough examination is required to clarify the distinctions and emphasize the advantages and disadvantages of each product.

3.7 Tools Used to Calculate LCA

There are many tools available to calculate product and process LCA. However, LCA is a data-intensive assessment largely dependent on available data and inputs. Several tools like BIM, etc. embedded with emission factors databases have been developed to improve efficiency and reduce the workload of EC assessment, as seen in Table 3.2, which lists available tools and datasets.

Apart from digital tools and databases, LCA can be done using manual calculations with the same input data and methods. The process and stages of an LCA calculation are discussed in upcoming chapters.

Table 3.2 Tools used to calculate LCA

No.	Software	Organization	Method	Phase	Source	Advantages	Disadvantages
1	Athena Impact Estimator	Athena Sustainable Materials Institute	Process	Cradle to cradle	[6]	– Simple and user-friendly – Covers cradle-to-cradle system	– Concrete mixture design options – Lack of regular updates
2	BEES	US	EI/O	Cradle to grave	[6]	– Simple to use	– Applicable only in North America – Concrete limited strength
3	eBalance	IKE Environmental technology CO, Ltd., China	Process	Cradle to gate	[6]	– High-quality life cycle database	– Only used in China with Chinese
4	EIO-LCA	Carnegie Mellon University, US	EI/O	Cradle to gate	[6]	– Impacts of different types of products through supply chain	– More suitable for universal products instead of single buildings
5	EQUER	IZUBA energies	EI/O	Cradle to grave	[6]	National database of France	– Recycling data is not detailed – Only used in France
7	ECO-BAT	Eco-Bat Technologies Ltd.	.Process	Cradle to grave	[6]	– Assessment is fast	Conception section not addressed
8	GaBi	PE International GmbH	Process	Cradle to grave	[6]	– Computation compatibility is superior	License required
9	OpenLCA	Gdelta	Process	Cradle to grave	[6]	– Variety of languages – Visualized results – Continuous updation	– Reliability of data not guaranteed
10	SimaPro	Pre-Sustainability	Process	Cradle to grave	[6]	– Detailed tool – Used in over 80 countries – Very flexible	– More time to use and learn the software

(continued)

Table 3.2 (continued)

No.	Software	Organization	Method	Phase	Source	Advantages	Disadvantages
11	Umberto	Ifu Hamburg	Process	Cradle to grave	[6]	– Applicable to users in Germany and other American countries	– Without reference data – Data import is strict
12	One-Click LCA	One Click LCA	Process	Cradle to cradle	N/A	– Simple and user-friendly – Accessible to all	– Limitations of databases available for emission factors
13	EC3	50+ industry leaders	Process	Cradle to gate	N/A	– Based on verified EPDs so more accurate	– Since it is EPD based, it is limited in terms of geography based on data available

References

1. Pallas G (2021) Life cycle-based sustainability standards and guidelines, May 11. Retrieved from Pre-sustainability: https://pre-sustainability.com/articles/lca-standards-and-guidelines/
2. World Green Building Council (2019) Bringing embodied carbon upfront. Ramboll and World GBC
3. Esau R, Jungclaus M, Olgyay V, Rempher A (2021) Reducing embodied carbon in buildings—low cost and high value opportunities. RMI
4. Weir M, Rempher A, Esau R (2023) Embodied carbon 101: building materials, March 27. Retrieved from RMI: https://rmi.org/embodied-carbon-101/
5. Huijbregts MAJ, Steinmann ZJN, Elshout PMF, Stem G, Verones F, Vieira M, Zijp M, Hollander A, van Zelm R. ReCiPe2016: a harmonized life cycle impact assessment method at midpoint and endpoint level (link is external). Inter J LCA. DOI: https://doi.org/10.1007/s11367-016-1246-y
6. Pan W, Teng Y (2021) A systematic investigation into the methodological variables of embodied carbon assessment of buildings. Renew Sustain Energy Rev 141, 110840. doi: https://doi.org/10.1016/j.rser.2021.110840

Chapter 4
Life Cycle Assessment Methodology

Life cycle assessment (LCA) can assist designers in determining the environmental repercussions of various designs by comparing structures, materials, and assembly. LCA is an iterative process, and this contributes to the consistency and comprehensiveness of the study.

4.1 LCA Stages

There are four stages of how the LCA is carried out:

1. **Goal and scope definition**—It defines the purpose, functional parameters, and boundaries of the study based on the LCA modules. Cradle to Gate (A1–A3), Cradle to Site (A1–A5), Cradle to Grave (A1–C4) as described in Fig. 2.2.
   It is important to define the functional unit, the system, and the limits of the analysis in the study.
2. **Life cycle inventory analysis**—This stage quantifies the inputs (such as energy and natural resources) and outputs (such as emissions to land, air, and water) and forms the core of the LCA study. It involves analyzing the flows into and out of the system defined for the study, including transport and material information, as well as replacement frequencies.
3. **Life cycle impact assessment**—This stage evaluates environmental impact in terms of standardized and key indicators such as Primary Energy, GWP, Ozone depletion, and others.
4. **Interpretation of results**—This stage involves critical review, determination of sensitivity, and result presentation [1] (Fig. 4.1).

A simple example describing the process is given in Fig. 4.2.

S. Mohan and Vijayalaxmi J., *Embodied and Operational Carbon in Buildings*,
SpringerBriefs in Architectural Design and Technology,
https://doi.org/10.1007/978-981-97-7187-5_4

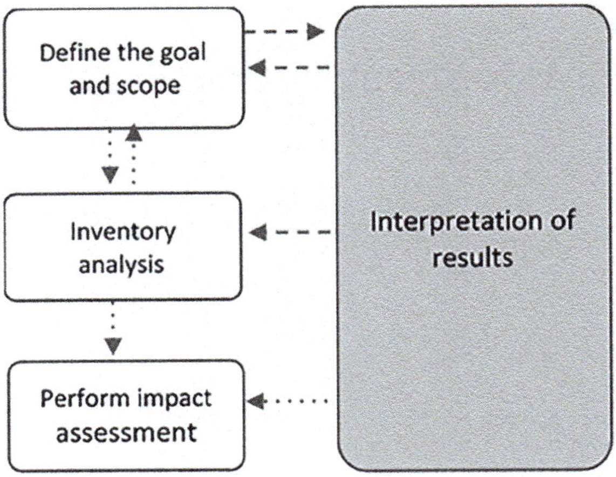

Fig. 4.1 Life cycle assessment of buildings—a practice guide [2]

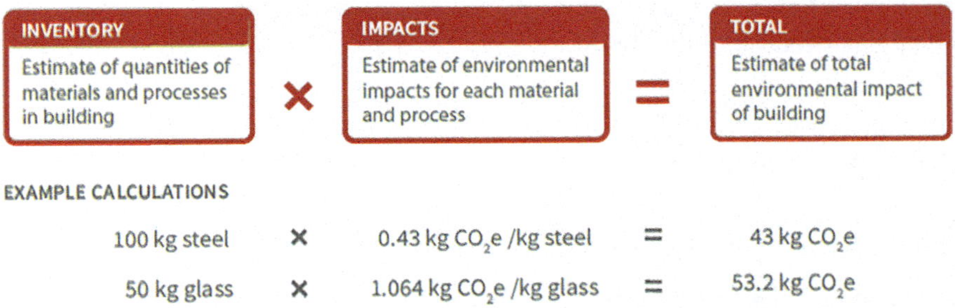

Fig. 4.2 Example of an LCA calculation

4.1.1 Objectives and Scope of the Life Cycle Analysis

The first stage in performing an LCA is to determine the assessment's goal and scope. This phase is critical because establishing why and how the LCA will be undertaken will assist in focusing efforts and ensuring that the work meets the aim.

Determination of the LCA goals, scope, and purpose for a specific decision system, such as whether to choose the most ecologically efficient material among accessible alternatives or to improve the material/product life cycle to improve environmental performance.

The following items form a part of the goal and scope definition:

- The functions of the system.
- The functional unit.
- The system to be studied.
- The system boundaries.
- Allocation procedures.

- Decision on impacts and measurements to be used.
- Data requirement.
- Assumptions and limitations.
- Data quality requirements.
- Any further critical review.

A few of these are elaborated below:

1. **A functional overview of the system**—LCA standards use a functional equivalent or a functional unit to characterize the subject of the assessment using objective criteria. The primary purpose of a functional unit is to provide a reference to which the inputs and outputs are related.
   When conducting LCA comparisons, it is critical to ensure that the options are equal in terms of a wide range of performance attributes over the course of an object's or building's whole life cycle and this necessitates the use of functional equivalency.
2. **Reference study period**—The reference study period is the time frame, usually defined in years often corresponds to the building lifespan. It is also known as the necessary service life.
   However, the actual building lifecycle is intrinsically unknown due to factors such as design quality, number of owners, and redevelopment possibilities. In this case, it can be determined in the following ways:

   (1) If a specific green building rating system is being employed, the rating system's mandated service life must be used as the LCA reference study period.
   (2) The model should be run twice: first with a lifespan of 50 years and again with a lifespan of 100 years.
   (3) Varying the building's lifespan and comparing the effect on the results helps determine dependency on the structure's lifespan.

3. **System boundary**

   (i) Building and site scopes explain the physical components of the building and project site that will be included.
   The system boundary defines the unit processes to be included in the system. The choice of these processes depends on goal and scope definition, intended application and audience, assumptions, data and cost constraints, and the cut-off criteria.
   (ii) Life cycle scope defines the life cycle stages that will be included.

   The setting of the system boundary, unit processes, flows and life cycle stages considering everything from acquisition of raw materials to production and use of fuels, electricity, and heat and up to manufacture, maintenance, and decommissioning of capital equipment.

4.1.2 Inventory Generation

A detailed inventory analysis must be performed to identify and quantify the environmental inputs (such as water, energy, and land), other resources, and outputs (such as byproducts) connected with a material throughout its life cycle. Each of these input-output categories may include simple materials, energy input, or some processes such as cement kilns or others, which must be broken down and analyzed similarly to the primary material life cycle processes.

- The inventory analysis contains the following main parameters.
- Collection of data.
- System boundary refinement based on data collected.
- Calculation of inputs and outputs.
- Data validation.
- Scrutinization of data based on system.
- Allocation can approximate as much of fundamental input-output relationships when dealing with complex multi-input and output processed or open loop methods (Fig. 4.3).

Following this analysis, a life cycle inventory should be developed, which includes data gathering for energy consumption, environmental discharge (gases, water, byproducts, or water), and resource consumption at each stage of the material life cycle.

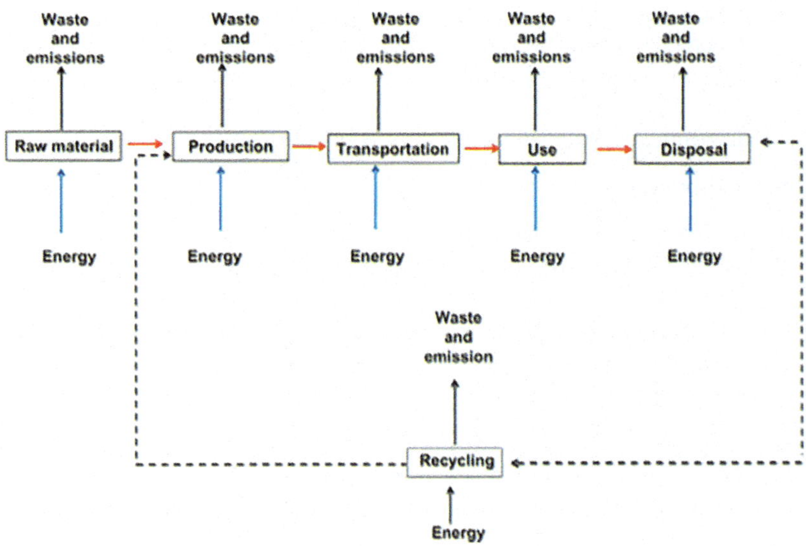

Fig. 4.3 Inventory analysis [3]

Data Collection

Data and information should be collected for:

- The amount of energy used at each step, the environmental impact, and any waste or byproducts produced at each step,
- The secondary material or processes needed for every phase of the life cycle of the primary material,
- The amount of energy needed to produce these secondary materials or carry out the procedures required to produce the primary material,
- The waste or byproducts from the secondary process as well as any environmental releases would contribute to the total,
- The entire amount of natural resources used for non-energy purposes at every point in the material life cycle, and
- The entire input cost at every point in the material's life cycle.

The inventory data will need to take into consideration data quality, data completeness, and geographic variability.

4.1.3 Environmental Impact Assessment on the Inventory Data

The impact assessment will evaluate the inputs and outputs in terms of their environmental impact. It contains the following processes:

- Category definition.
- Classification.
- Categorization.
- Valuation and Weighting.

The indicators for environmental impact can be qualitative or quantitative and have been discussed in Chap. 3 to understand how LCA can be leveraged. It is preferable to use only one source of data inside an LCA because data from multiple LCIs are not always comparable. However, if numerous sources of data were used, they must be explicitly disclosed.

A number of key considerations while selecting the impact category are:

- Completeness.
- Practicality.
- Independence.

Valuation and Weighting

After an initial review of the LCA results, the division of the environmental impacts can be done by certain critical parameters, such as building component or material type, and/or per life cycle stage. A visualization of the results at this stage can further aid the assessment. It can determine which building components, materials, and life cycle phases have the most impact.

There are various methods to do valuation and weighting based on different principles and the one which aligns with the scope of the study should be considered.

As discussed, the five most often monitored environmental effect categories in LCA are:

a. Global warming potential (GWP),
b. Ozone depletion potential (ODP),
c. Eutrophication potential (EP),
d. Acidification potential (AP), and
e. Smog formation potential (SFP).

4.1.4 Interpretation of Results

Three distinct activities occur as a part of the interpretation process, and these include:

- Identification of significant impacts.
- Evaluation.
- Conclusion and recommendations.

It is one of the least developed portions of the international standard and is expected to be revised soon.

If the results of the inventory analysis or the impact assessment are found not to fall short of the goal and scope definition, the inventory analysis must be improved by either revising the boundaries, changing the data sources, or further data collection, etc. This is an iterative process and must be carried out till the goal and scope are fulfilled (Fig. 4.4).

4.2 Integration and Check of Results

1. **Sensitivity Analysis**: A sensitivity or uncertainty analysis can assist in understanding the importance of these variables.
   A sensitivity analysis helps in determining which variables in LCA have the most impact on the outcomes. This variable could be a physical component like a material, assembly, or energy source, or it could be an assumption like the projected lifespan or reference study period.
2. **Completeness check**: A completeness check to analyze if the inventory used is unavailable or incomplete. Necessary resources should be used to improve the data set if this is the case.
3. **Consistency check**: A consistency check is qualitative. The methodology and sources should be used consistently across the study. The variations in regions or temporal differences, and other methods should be evaluated.

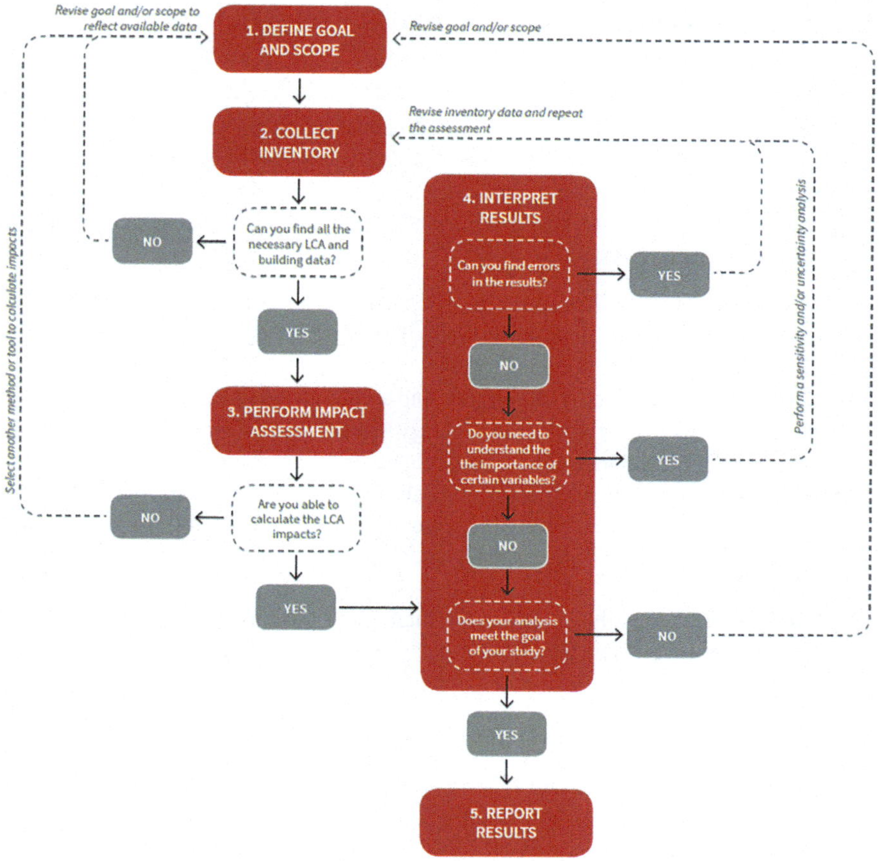

Fig. 4.4 Life cycle assessment [4]

This can be supplemented by data from an uncertainty analysis or data quality assessment.

Uncertainty analysis: An uncertainty analysis determines which components of the LCA are the most uncertain. This is significant since it influences the effectiveness of the results.

Data quality assessment: This is a series of qualitative checks to ensure the data collected in the inventory passes the quality standards and is not incomplete or inaccurate.

4.3 Methodological Variables in LCA

Further to the assessment, some key variables can directly affect the outcome of the assessment. A study by Pan and Teng gave rise to the discussed methodological variables involved in LCA calculation. Further study by Dixit et al. showed the reliability of variables on certain parameters as shown in Fig. 4.5.

Some of the key variables that fall into these categories and that affect Life Cycle Assessment are as follows:

1. **Material selection**: The selection of construction materials significantly influences LCA. Material extraction, manufacturing processes, and transportation contribute to environmental impacts.
2. **Construction process**: The methods and technologies used during the construction phase play a crucial role. Energy consumption, waste generation, and emissions during construction affect the overall life cycle.
3. **Building operation and maintenance**: Energy consumption, water use, and maintenance practices during the operation phase contribute to the environmental impact. Effective systems and regular maintenance can mitigate the negative effects.
4. **Energy efficiency**: the energy performance of a building is a critical factor. The implementation of energy-efficient technologies and renewable energy sources can have a positive impact on LCA.
5. **Transportation**: The transportation of materials to the site, as well as the daily movements of residents, contributes to the carbon footprint. Proximity to suppliers and public transport options can influence this aspect.

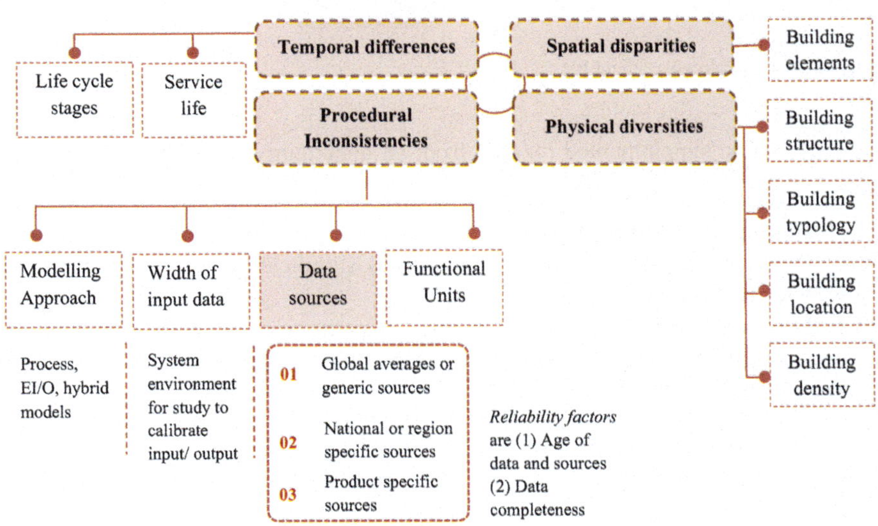

Fig. 4.5 System framework for methodological variables of LCA

6. **Demolition and end of life**: How a building is demolished and the disposal of construction waste affects the end-of-life phase. Sustainable demolition practices and material recycling can reduce environmental impact.
7. **Location and choice of site**: The geographical location of a building influences factors such as the climate, the availability of energy, and the availability of sustainable resources. The right choice of the site can optimize the energy efficiency.
8. **Occupancy patterns**: How occupants use and interact with the building influences energy and resource consumption. The efficient use of spaces and responsible behavior can contribute to a more sustainable life cycle.
9. **Adaptability and flexibility**: Buildings designed with adaptability in mind can potentially extend their life cycle, reducing the need for new construction. This longevity has a positive impact on overall environmental performance.
10. **Water Efficiency**: Sustainable water management practices, including water-efficient fixtures and landscaping, help reduce environmental impact during the operational phase.
11. **Innovation and technology**: Integration of innovative technologies and sustainable construction practices can positively influence LCA, reducing resource consumption and improving overall efficiency.

4.4 Summary

Energy-efficient design and construction practices play a crucial role in reducing energy consumption and carbon emissions in public buildings.

Overall, LCA serves as a valuable tool for assessing and improving the environmental performance of buildings, supporting sustainable decision-making, and promoting the adoption of environmentally responsible practices in the building industry.

The potential development of LCA in the next few years include:

1. LCA will form an intrinsic part of the environment management tools, with increased applications for corporate strategy.
2. Demand for life cycle information from suppliers and manufacturers is on the rise and will continue to increase. The implementation of the Carbon Border Adjustment Mechanism (CBAM) in 2026, is likely to further aid this monitoring and verification of production and process-based emissions.
3. Development of more widely accepted standards and methodologies. This may or may not vary on a regional basis.
4. Market pressures will push for greater benchmarking and industry averages.
5. There should be more commonality and greater availability of data.
6. Pressure for green products and eco-labeling schemes may increase as knowledge of LCA methodology increases in the industry.

Currently, sustainable development is evolving exponentially in importance and is integrated into national and international agendas. This requires rapid improvements in terms of energy and material efficiency. This can be done more effectively across the life cycle of the process. The importance of LCA is therefore only going to grow further and the industry should be ready to tackle the challenge.

The applicability of the LCA study is therefore of immense importance and detailed further in upcoming chapters.

References

1. Muralikrishna IV, Manickam V (2017) Chapter Five—Life cycle assessment. In Muralikrishna IV, Manickam V (eds) Environmental management. Elsevier, pp 57–75
2. ISO 14044:2006 (2006) Environmental management: Life cycle assessment—Requirements and guidelines. ISO, 2006
3. ul Islam S, Kumar A (2019) Life cycle assessment of rubber dam. In: Plastics design library, hydraulic rubber dam, William Andrew Publishing, pp 99–122. ISBN 9780128122105. https://doi.org/10.1016/B978-0-12-812210-5.00008-0
4. Moazzen N, Karagüler ME, Ashrafian T (2019) Life cycle energy assessment of a school building under envelope retrofit: An approach towards environmental impact reduction. E3S Web of Conferences, 111, 03028. https://doi.org/10.1051/e3sconf/201911103028

Chapter 5
Strategies to Decarbonize

5.1 Introduction

Life cycle assessment (LCA) can be used in design and construction as a powerful, science-based tool to decarbonize the built environment and drive progress on climate goals.

There are a number of factors that affect the life cycle of a building such as transport, geometry of the building, or manufacture of a material. These factors, if effectively handled at the design and construction stage, can significantly aid reduction of carbon emissions.

5.2 Strategies for Carbon and Energy Reductions

The strategies to reduce carbon emissions can be integrated at an early stage to have more impact on reductions. A summary of a few of the strategies to reduce embodied carbon and energy is described (Table 5.1).

5.3 Studies of Carbon Reduction Strategies Used in Buildings

In order to assess the impact of carbon reduction strategies and how this is addressed through rating systems in India, a series of green building case studies are studied.

Sixteen green buildings, which are rated by the Indian Green Building Council (IGBC), Green Rating for Integrated Habitat Assessment (GRIHA), and Leadership In Energy and Environmental Design (LEED) are considered for the study as listed in Table 5.2.

S. Mohan and Vijayalaxmi J., *Embodied and Operational Carbon in Buildings*, SpringerBriefs in Architectural Design and Technology,
https://doi.org/10.1007/978-981-97-7187-5_5

Table 5.1 Strategies for embodied and operational carbon reduction

Building stage	Strategy
Embodied carbon reduction strategies	
Feasibility	Retrofit and reuse existing buildings
Concept design	Reduce floor area
Detailed design	Reduce below-grade construction
Concept design	Design lightweight efficient structures
Detailed design	Use tools to optimize envelope design
Concept design	Select materials for structure, envelope, insulation, and finishes
Detailed design	Selection of low-carbon MEP systems
Concept design	Select salvaged or refurbished materials
Concept design	Design for disassembly
Tender stage	Integrate EPDs and GWP into specifications for procurement strategies
Detailed design	Optimize concrete specification and mix design
Tender stage	Evaluate cost and carbon in bids
Operational carbon reduction strategies	
Use stage	Efficient building operations
Concept design	Building envelope materials
Detailed design	HVAC systems
Detailed design	Lighting and energy use reductions
Detailed design	Renewable energy systems
Use stage	Building management systems
Use stage	Energy-efficient appliances
Use stage	Smart technology

All the studied buildings are office buildings in India and are considered with a life of 40 years.

The buildings studied are of varying scales and specifications. However, using scale as a metric for comparison, the buildings are evaluated on the basis of their embodied and operational emissions.

The scales considered are based on the building sizes indicated in Table 5.2:

1. Buildings 1–6—Area between 15000 and 30000 m^2
2. Buildings 7–11—Area >30000 m^2
3. Buildings 12–16—Area <15000 m^2

The methodology followed for the study used this basic building information from Table 5.2 in the following tools.

Table 5.2 Buildings considered in the review [1, 2]

Building	Climate	Area (m^2)	Height and structure	Certification	Conditioned space (%)
Building 1	Warm and humid	30235	6-storey concrete structure	GRIHA—Five star	None
Building 2	Moderate	22860	4-storey concrete structure	LEED—Platinum GRIHA—Five star	100
Building 3	Warm and humid	24443	5-storey concrete structure	IGBC—Net Zero Rating and BEE 5-star rating	65
Building 4	Composite	23500	9-storey concrete structure	IGBC—Platinum Rating	100
Building 5	Composite	16260	8-storey concrete structure	LEED—Platinum	65
Building 6	Warm and humid	22296	6-storey concrete structure	GRIHA—Five star and LEED—Platinum	70
Building 7	Composite	37874	6-storey concrete structure	GRIHA—Five star	65
Building 8	Composite	48000	4-storey concrete structure	LEED Platinum	70
Building 9	Composite	31400	7-storey concrete structure	GRIHA—Five star and LEED Platinum	85
Building 10	Moderate	70865	3-storey concrete structure	GRIHA—Five star	57
Building 11	Moderate	72465	5-storey concrete structure	IGBC—Platinum and LEED—Platinum	N/A
Building 12	Composite	8635	Concrete structure	GRIHA—Exemplary Award	55
Building 13	Moderate	3205	5-storey concrete structure	GRIHA—Four star	28
Building 14	Hot and dry	5100	3-storey concrete structure	GRIHA—Five star	10

(continued)

Table 5.2 (continued)

Building	Climate	Area (m^2)	Height and structure	Certification	Conditioned space (%)
Building 15	Composite	3720	3-storey concrete structure	LEED Platinum	N/A
Building 16	Warm and humid	2323	3-storey concrete structure	LEED Platinum	N/A

For embodied carbon: The building was modeled using 'Carbon designer' and 'One Click LCA' for an impact of the embodied carbon emissions. This was also compared to the 'embodied carbon calculator' EC3 for A1–A3 carbon or upfront carbon emissions.

For operational carbon: The operational energy use (B6) of the building is converted to operational carbon emissions, using the carbon emission factor for electricity in India.

While this may not give an accurate picture of whole life carbon of each building, on a comparative scale, the carbon emissions and strategies used may inform the applicability of certain guidelines for embodied and operational carbon reductions.

The results of these studies are described in Fig. 5.1.

This figure describes a few of the buildings studied for carbon emissions across all scales. It indicates the carbon emissions as a bar chart, followed by embedded emissions and operational energy aggregated as lines across the chart across all scales of buildings in Table 5.2.

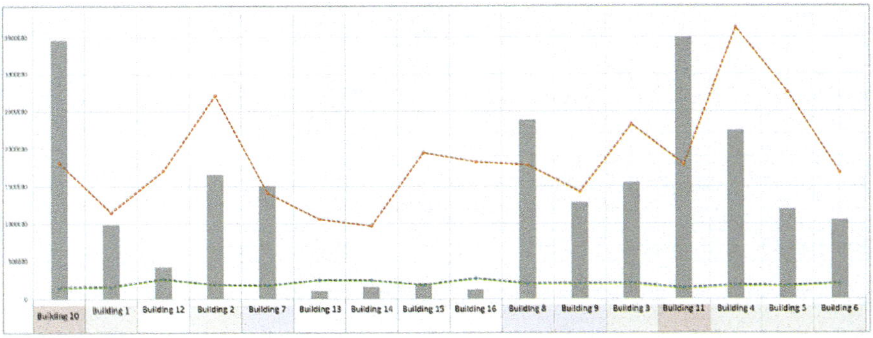

Fig. 5.1 Indication of embodied and operational carbon of buildings

5.3.1 *Comparison of Buildings of Scale Between 15000 and 30000 m²*

A summary and comparative study of the carbon emissions from office buildings with a floor area between 15000 and 30000 m^2 is discussed in this section.

The metric used, is $kgCO_2/m^2/year$, to provide more comparability to the calculated embodied and operational emissions (Table 5.3).

Strategies used for carbon reductions in studies:

- Buildings 2 and 4 have a similar carbon footprint and air conditioning space; however, the height and area are variable and comparable.
- The strategies for Building Materials used here are:
  - 83% of materials in Building 2 are low carbon comprising Autoclaved aerated concrete AAC blocks with 65% fly ash content.
  - In Building 4, cavity walls and reflected insulated roofs are used, which have lower carbon than in Building 3. Along with this, roof gardens and double-glazed windows are used.
- Building 6 has embodied carbon on the higher side and operational carbon on the lower side of the scale. This is due to the use of the radiant cooling system for air conditioning of the building. Building 3 has a wind funnel integrated into courtyard planning, leading to a reduction of operational carbon emissions through passive design (Table 5.4).

Table 5.3 Summary of carbon emissions from buildings under 30000 m^2 [1, 2]

Metrics	Units	Building 1	Building 2	Building 3	Building 4	Building 5	Building 6
		LEED	LEED	GRIHA	IGBC	IGBC	LEED
EC	$kgCO_2/m^2/year$	3.95	4.625	5.175	4.475	4.325	4.975
OC		28.7	67.65	58	90.72	68.85	41.99
Area	m^2	30235	22860	24443	23500	16260	22296
CF	$kgCO_2/m^2/year$	32.65	72.27	63.24	95.19	73.175	46.97

Table 5.4 Summary of carbon emissions for buildings of scale >30000 m^2 [1, 2]

Metrics	Units	Building 7	Building 8	Building 9	Building 10	Building 11
		LEED	LEED	GRIHA	GRIHA	IGBC
EC	$kgCO_2/m^2/year$	3.6	4.4	5.075	5.075	3.525
OC		35.24	44.55	35.64	45.24	44.77
Area	m^2	30235	22860	24443	23500	16260
CF	$kgCO_2/m^2/year$	**39.63**	**49.62**	**40.71**	**48.85**	**48.3**

5.3.2 *Comparison of Buildings of Scale >30000 m²*

Strategies used for carbon reductions in studies:

- Buildings 7 and 9 are compared based on similar areas and carbon emissions. Building 7 uses pozzolana Portland cement with 32% fly ash content in all plasters and mortars. Building 9 uses AAC block in masonry walls, stone, and Jaali walls (lattice screen wall). Yet building 7 has a lower embodied carbon value indicating that replacement of structural material in specification or in reduction of height is crucial for embodied emissions.
- Building 11 is 40% more efficient than the ASHRAE standard through envelope material and double-glazed windows strategies. Additionally, 10% of the total construction material used was recycled including steel and glass. 41% of the total project material by cost was manufactured regionally.

5.3.3 *Commonly Used Strategies and Their Impacts from Case Studies*

Using the studies from Sects. 5.3.1 and 5.3.2, a trend that may be observed is that the increasing proportion of embodied emissions can be attributed to efforts to reduce operational emissions. For example, using photovoltaics or adding insulation materials, double glazing, etc.

While some strategies do affect embodied emissions such as material selections or retrofit decisions, carbon reduction cannot be successful with only targeting operational energy or thermal comfort. The importance of embodied emissions when making decisions regarding emissions reduction strategies should therefore be acknowledged uniquely and treated with utmost seriousness. This is particularly relevant in terms of guidelines, standards, or voluntary certifications like the green rating systems.

The strategies used in the studies and their connection to embodied and operational carbon are summarized in Table 5.5.

Other international examples include strategies like using glass facades in building designs, green roofs, walls, etc., which also have considerations in rating systems like IGBC and GRIHA.

It is seen from the case studies, that replacing the structural materials and use of AAC blocks has some of the lowest embodied carbon values, followed by the use of cavity walls and radiant cooling systems. However, it should be noted, that strategies determination depends on geography, climate, and scale of the building.

Though they are context specific, they can be broadly aggregated into design strategies over the proven impact on carbon emissions.

Table 5.5 Summary of commonly used strategies in studies and if they are addressed in rating systems

Strategies from studied buildings	Relation to OC	Case studies		EC range	How it is addressed		
		Area		$kgCO_2e/m^2/yr$	IGBC	GRIHA	LEED–USGBC
1. Use of AAC blocks as an envelope material Limestone: Cement: Fly Ash: Aluminium powder (8:33:59:0.07)	Impacts heat transfer and thermal performance which has a direct relationship to OC and EC	<8000 m^2	4	6.1–6.7	At least 10% the cost of the building is met by environmental criteria (recycled)At least 7.5% by cost of salvaged materialAt least 30% by cost recycled material	Concrete blocks with minimum 40% FA contentReduction of LCA of the building super-structure 15%	Demonstrate 5–20% reductions in structural and enclosure GWP with reuse/salvaged material using cradle to grave LCA
		15–30 k m^2	4	4.3-4.6			
		30–50 k m^2	1	5.08			
2. Use of cavity walls for envelope	Impacts heat transfer and thermal performance	<8000 m^2		4.78–4.98	At least 10% the cost of the building is met by environmental criteria (recycled)	Does not include cavity walls in the list of alternative materials	Show 5-20% reductions in structural/enclosure GWP with cradle to grave LCA
		15–30 k m^2	1	6.1			
3. Radiant cooling systems	Directly impacts operating energy and emissions	< 8000 m^2		4.55	EPI for an office building in composite climate is 110 KWh/m^2/year	EPI ratio must be less than or equal to 1 compared to the EPI benchmark BEE building	Performance Cost Index target depicts the improvement using greenhouse gas targets
		30–50 k m^2	1	4.98–5.08			

(continued)

Table 5.5 (continued)

Strategies from studied buildings	Relation to OC	Case studies		EC range	How it is addressed		
		Area		$kgCO_2e/m^2/yr$	IGBC	GRIHA	LEED–USGBC
4. Envelope retrofit	Directly impacts embodied carbon by not changing any new primary materials	<8000 m^2	–	–	Retention of at least 75% of the structural elements and at least 50% of non-structural elements (area) of the existing building	Building can be retrofit with certain environmental criterial like energy and materials as specified.	Structural material retained up to 75% at least 30% of all non-structural elements
		15 $m^2$30 k m^2	1	5.18			
5. Replace structural material	Directly impacts embodied carbon and operational carbon, more if the structure is exposed	<8000 m^2	3	4.98–5.08	All of the criteria address enclosure and structure together separated by types of material	Structural material by weight of cement 30% replaced by waste material or 5% aggregate recycled	Demonstrate 5–20% reductions in structural GWP with reuse/salvaged material using cradle to grave LCA
		15–30 k m^2	5	4.3–4.6			

5.4 Guidelines for Embodied and Operation Carbon Emissions Reduction

The development of design-based guidelines for embodied and operation carbon emissions reduction is based on the above sections, Sects. 5.2 and 5.3. These guidelines should be flexible for possible use in the context of green building rating systems or as design guidelines for architects and design professionals.

The broad categories synthesized from the studies and building examples are:

1. Passive design
   (a) Structural systems
   (b) Alternative materials
2. HVAC Systems

These categories are intended to target embodied and operational emissions at the design stage for a proposed building.

5.4.1 Passive Design

Energy required by the building sector in India accounts for 33% of the national energy use. Climate-responsive design can drive the building sector in India toward a low-energy future since including passive design approaches and materials can decrease cooling by 60–90% for new buildings and 50–90% for retrofits [3].

Different passive design strategies are applicable to various climate zones. These are:

1. Building layout (position, orientation, and overshadowing).
2. Envelope thermophysics (fabric, ventilation, and shading).
3. Building geometry.
4. Airtightness and infiltration.

Passive design strategies are not granted the same credit value as a traditional whole-building energy simulation approach and should be accurately defined and more comprehensively addressed in current guidelines and rating systems. For example, BREEAM is a good example which addresses passive design such that an appropriate degree of thermal comfort must be secured before passive credits can be claimed [4].

The passive design strategies which directly impact embodied carbon such as fabric-based design decisions, are detailed in this book. Other passive strategies, like form, orientation, or shading, though impactful at the design stage are numerous and heavily dependent on context. Therefore, they are not addressed in detail.

5.4.1.1 Target Structural Systems

As discussed in Chap. 4, structural systems and materials have a large impact on embodied emissions. However, using this variable, carbon emissions can be reduced by almost 50% [5].

The parameters of structural systems that broadly impact carbon are:

- Volume or Quantity of the material.
- Type of structural system.
- Type of structural material used.

In the context of green building rating systems, two important compliance criteria considerations include,

1. The object of assessment—For example, which materials, components, operating energy, water systems, etc. are to be included and
2. The assessment system boundary—For example, which life cycle processes of the object need to be considered.

Examples of mitigation measures based on these parameters include the following:

1. In the case of an existing building, retrofit and reuse can be considered for 75% of the building.
2. Replacing at least 20% of the cement in concrete with alternate, bio-based, or recycled aggregates.

This parameter targets structural systems and their material decisions. The type of structural system is an important parameter to be considered based on the stability and efficiency of the system.

Structural System-Based Carbon Reduction Measures

Mitigation measures for structural systems are broadly divided into two categories as seen in Fig. 5.2.

- Design-based changes—Form optimization, Type of system, etc.
- Material-based changes—Type of structural material, specification, mix proportion, etc.

Guidelines developed

The following guidelines are developed from the study of variables from Table 5.1 and case studies discussed in Chap. 3 (Table 5.6).

5.4.1.2 Alternative Materials

Alternate materials play a major role in addressing embodied carbon. This is also due to the fact that the volume of the enclosure material used in proposed developments is very high.

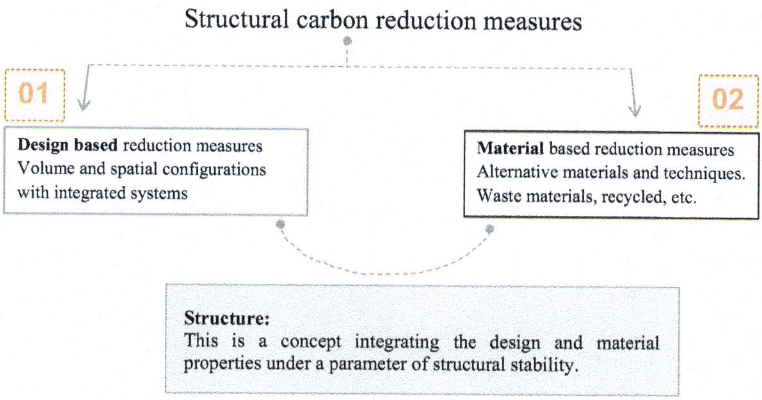

Fig. 5.2 Structural parameters for design integration

The use of alternative building materials like natural materials, biomaterials and salvaged materials that have lesser embodied carbon compared to conventional building materials can aid in reducing carbon emissions.

In the context of the green rating system, alternative materials are flexibly and articulately addressed, with thresholds defined. The two key parameters to address these materials are

1. The typology of alternative materials used (e.g., biomaterials, waste materials, natural materials, etc., which is dependent on the local context of the building),
2. The threshold percentage of replacement or use of the described material in the building.

Examples of mitigation measures based on these parameters are as follows:

- For new construction, at least 10 or 20 materials need to have EPD from the manufacturers and from at least 5 different manufacturers. While this does not directly address carbon emissions, in combination with the Materials and Resources LCA credit, these criteria can result in substantial carbon savings.

Guidelines developed

The material savings can be brought into effect by either,

(1) Using low-carbon materials or
(2) Reducing the volume of carbon-intensive material used.

Based on these parameters, guidelines developed are discussed in Table 5.7 with case studies, to describe the potential impact of the strategies' system configurations.

Table 5.6 Guidelines for structural systems parameter

Parameters	Guideline	Type	Stage	Consideration	Application examples	Impact of case studies
Design-based measure	Type of structural system used	New buildings	Concept stage	Available local material and craftsmanship (context) and structural viability of the system	Steel-based structural system (with 100% recycled content)	5–15% (Chap. 1)
					Precast concrete framed structure	
Material-based measure	Structural material used	New buildings	Concept stage	Locally produced agricultural and industrial waste can be used for cement replacements in green concrete	Agro-based green concrete mixes (RHA, BLA, etc)	9-80% [6]
					Use of precast concrete for the structure (columns and beams)	
				Mix proportions of additive without compromising structural properties	Industrial waste-based concrete (Fly ash, GGBS, etc)	up to 60% [7]
					Use of bamboo in slabs	up to 50% [8]

Table 5.7 Guidelines for alternative material parameters

Parameters	Guideline	Type	Stage	Consideration	Application examples	Impact of case studies
Alternative materials	Exterior enclosure low-carbon materials	New buildings	Concept stage and Design development stage	Available local material and craftsmanship (context) and stability of the materials	Bio-based materials.	19–75% [9]
					Natural materials	12–66% [10]
					Waste based materials	up to 30% [11]
					Salvaged materials	N/A

Table 5.8 Guidelines for HVAC system parameters

Parameters	Guideline	Type	Stage	Consideration	Application examples	Impact of case studies
Systems configuration	HVAC	New buildings	Concept stage and design development stage	Type of system	Radiant cooling	10–70%
					Water-based chiller systems	

5.4.2 *HVAC System*

The HVAC systems play a major role in the carbon reduction strategies to effect significant savings. HVAC has the maximum share in the energy consumption of a building, which is almost 45% of the building's load. Any change or reduction in HVAC load, using management systems, smart operating, and metering, efficient co-efficient of performance and system type, will cause a significant reduction in energy consumption, which in turn affects operational carbon emissions. However, HVAC has an important role to play in the thermal comfort of the user and hence, this must be evaluated and considered as well while making changes to reduce carbon emissions.

In current rating systems, optimized energy performance and energy efficiency are addressed, with a good focus on systems and configurations.

For example, LEED places a large emphasis on the marking for the Energy component, and out of 17 credits under Energy and Atmosphere, 15 are aimed at energy reduction or renewables, with 3 prerequisites.

Guidelines developed

For energy performance, some of the key considerations of systems include,

1. Type of energy-efficient system used like water cooled chiller plant or DOAS.
2. Efficiency of the system.
3. Refrigerant used.

The guidelines developed are largely in line with rating systems and are described in Table 5.8.

5.5 Summary

Currently, LCA is typically separate from the design process, usually incorporated after detailed design or in most cases, after the construction of the building. The necessity to create a zero carbon built environment implies that LCA integration in the early stage of design needs to be clearly addressed. This can be done using green building rating systems.

The use of a green building rating system can be directed into two major concerns:

1. To ensure the rating system addresses all the stages of building development right from the initial stage.
2. To target and reduce carbon emissions of buildings—with both embodied and operational sectors.

The system should target wholistic sustainable development, with a focus on the carbon aspects of building sustainability. The design stage mitigation measures could include reduced volume of materials, use of low carbon or reused and recycled materials, design for longevity and adaptability, retrofitting existing buildings, and education and awareness about embodied carbon. On the operational side, strategies include using energy-efficient appliances, optimizing façade, renewable energy sources, efficient building monitoring systems, and others. Not all of them can be addressed comprehensively at a design stage.

At the design stage, these mitigation measures are broadly divided into (1) passive design and (2) operational decisions. Passive design addresses form, orientation, ventilation, shading, material use, and others at a concept stage, while Operational decisions include decisions like the use of automation or integration of PV, type of HVAC system used, and others. This has been aggregated with the broad guidelines which are estimated to have maximum carbon impact.

These guidelines can be applied at a small-scale designer level or can be scaled up for inclusion in rating systems and standards, though this would include collaboration between many stakeholders and specialists.

In this context, a large number of buildings within the Indian GBRS have been studied to assess the impact of these strategies in their guidelines. It is found that almost all the rating systems include the strategic criteria discussed, moving toward all-inclusive sustainable development. However, the carbon emissions have not been quantified and no targets have been set. Establishing these quantification, methods, tools, and data lends an aspect of practicality to the current theoretical adoption of LCA in the Indian green building rating systems.

References

1. GRIHA (n.d.) Case studies. Retrieved from GRIHA India: https://www.grihaindia.org/case-study
2. Net Zero Energy Buildings (NZEB) (n.d.) Detailed case studies. Retrieved from NZEB: https://nzeb.in/case-studies/detailed-case-studies-2/
3. Passe U, Battaglia F (2015) Designing spaces for natural ventilation—an architect's guide. Routledge
4. Chen X, Yang H, Lu L (2015) A comprehensive review on passive design approaches in green building rating tools. Renew Sustain Energy Rev 1425–1436
5. RICS (2012) Methodology to calculate embodied carbon of materials. RICS

6. Blesson S, Rao A (2023) Agro-industrial-based wastes as supplementary cementitious or alkali-activated binder material: a comprehensive review. Innovative Infrastructure Solutions
7. Jiang P, Zhao D, Jin C, Ye S, Luan C, Tufail RF (2024) Compressive strength prediction and low-carbon optimization of fly ash geopolymer concrete based on big data and ensemble learning. PLoS ONE 19(9):e0310422. https://doi.org/10.1371/journal.pone.0310422
8. Sankannavar SA, Ravindra R, Basutkar SM (2024) Flexural strength of concrete slabs reinforced with bamboo of different knots spacing. Asian J Civ Eng 25:3495–3506. https://doi.org/10.1007/s42107-024-00992-3
9. Zuiderveen E, Kuipers K, Caldeira C, Hanssen SV, van der Hulst M, de Jonge MM … Huijbregts M (2023) The potential of emerging bio-based products to reduce environmental impacts. Nat Commun 14
10. Nanz L, Rauch M, Honermann T, Auer T (2019) Impacts on the embodied energy of rammed earth facades during production and construction stages. J Facade Des Eng 7(1):75–88
11. Chen L, Huang L, Hua J et al (2023) Green construction for low-carbon cities: a review. Environ Chem Lett 21:1627–1657. https://doi.org/10.1007/s10311-022-01544-4

Chapter 6
Embodied and Operational Carbon—Case of a Building

The methodology to calculate embodied and operational carbon emissions for a proposed building by the life cycle assessment is discussed in this chapter.

LCA follows the stages described in Chap. 4. The International Organization for Standardization (ISO) has developed a series of international standards to cover LCA such as ISO 14040 (LCA-Principals and guidelines), ISO 14041 (LCA-Life Inventory Analysis), ISO 14042 (LCA-Impact Assessment), and ISO 14043 (LCA-Interpretation).

The standards utilize a set of environmental indicators to quantify the impact of a proposed building. For example: EN15084 standard for the construction sector defines 15 environmental impact categories. Some of the indicators are:

1. Damage to Human Health: It covers particulate matter, climate change, toxicity, etc.,
2. Damage to the Ecosystem: It includes marine pollution, acidification, agricultural land occupation, etc.,
3. Resource Depletion: It includes water, mineral, and fossil fuel depletion.

Methodology and tools studied

As discussed in Chap. 3, the LCA study involves four stages. The tools and assumptions for the study are outlined below.

6.1 Goal and Scope Definition

The establishment of the objectives and the scope of the study are defined in this phase. The determination of scope, key areas of focus, assumptions, and exclusions are included here.

S. Mohan and Vijayalaxmi J., *Embodied and Operational Carbon in Buildings*, SpringerBriefs in Architectural Design and Technology,
https://doi.org/10.1007/978-981-97-7187-5_6

In this study:

The objective of this LCA study is to assess the impact of the design guidelines for carbon reduction developed in Chap. 4.

It is primarily a comparative analysis to identify the most impactful strategy and examine how it is addressed in green rating systems in India.

For this study, the following are considered:

Impact considerations: As it is intended to assess carbon reduction; the primary impact indicator is the Global Warming Potential or GWP of the building.

Building considerations: A baseline study is required to establish carbon emissions comparison and impact.

Data sources: The sources of data required, particularly for carbon emission factors, building material and operation data, and tools required to carry out the analysis.

System boundary: The life cycle boundary study conducted here is a **cradle-to-gate study** for embodied carbon emissions. This is because carbon emissions from A1 to A3 of the life cycle account for around 50% of the embodied carbon emissions of a building [1]. Operational carbon emissions from B6 account for more than 90% of operational emissions and are included in the system boundary.

Functional unit: The metrics used for embodied carbon is $kgCO_2/m^2$ and the operational carbon metrics are dependent on operational energy and use the units $kgCO_2/m^2$/year.

On the assumption that the life of the building is 60 years, a comparable metric of the format $kgCO_2/m^2$/year is used for the comparison of carbon emissions between various iterations. In this book, it is considered equivalent to whole life carbon.

6.2 Inventory Analysis

This stage encompasses all the inputs and outputs tied to the building. These inputs may be raw materials and energy, while outputs could be waste, water, and pollutants.

The first step is to identify the required data defined by the goal and scope definition in Chap. 4. The next step is the data collection phase. This is complex due to lack of availability or inconsistent sources and is also time consuming.

In this study, the Source of carbon data is as follows:

For embodied emissions—The materials used in the building are defined based on the rating systems and the associated carbon emission factor in India is tracked. The source database used is the IFC EDGE database for common materials in India. Where data is not available, the ICE global database or referenced research papers are used.

For operational emissions—Information about the systems, occupancy schedules, materials, etc. is input in a 3D model and simulated for operational energy use. A carbon factor for electricity is applied to get the associated carbon emissions of operation.

The building data: The data and information of the building is collected in the following way:

The drawings of the baseline study are laid out and the 3D form is pulled up. This is used for volumetric calculations. All iterations are applied over this, with no change to the essence of the building, except the parameter studied—for example, envelope material.

6.3 Impact Assessment

The impact assessment involves transforming the raw data into measurable impacts. This utilizes the life cycle database and information collected and quantifies the impact of indicators.

In this study:

The impact assessment criteria considered in this study deal with embodied carbon emissions, so the environmental indicator considered is global warming potential (GWP). The units used here are $kgCO_2e$.

6.4 Interpretation of Results

The results are interpreted to validate the conclusions of the impact assessments. It involves a completeness, consistency, and sensitivity check along with conclusions and recommendations.

6.5 Development of the Baseline Building

Criteria for development: The selected site should be in line with major Indian rating systems, particularly GRIHA and IGBC. The typology identified is an office building of a relevant scale and a regular occupancy schedule.

Hyderabad holds the record to be the most livable city in the country as per the Mercer Quality of Living Index. Further, it is a growing commercial center of the country, indicating an ideal location for the proposed building [2] (Fig. 6.1).

The development of the baseline building is inspired by Infosys Pocharam, which is an office campus, located in Hyderabad. It is a green-rated building, compliant with GRIHA and LEED rating systems, and has around 24,000 m^2 of built-up area. It is designed to minimize impacts on the environment, while providing a healthy environment for occupants as in Fig. 6.2.

The campus itself is spread on 447 acres of land just on the boundary of the city and is represented as seen in Fig. 6.3.

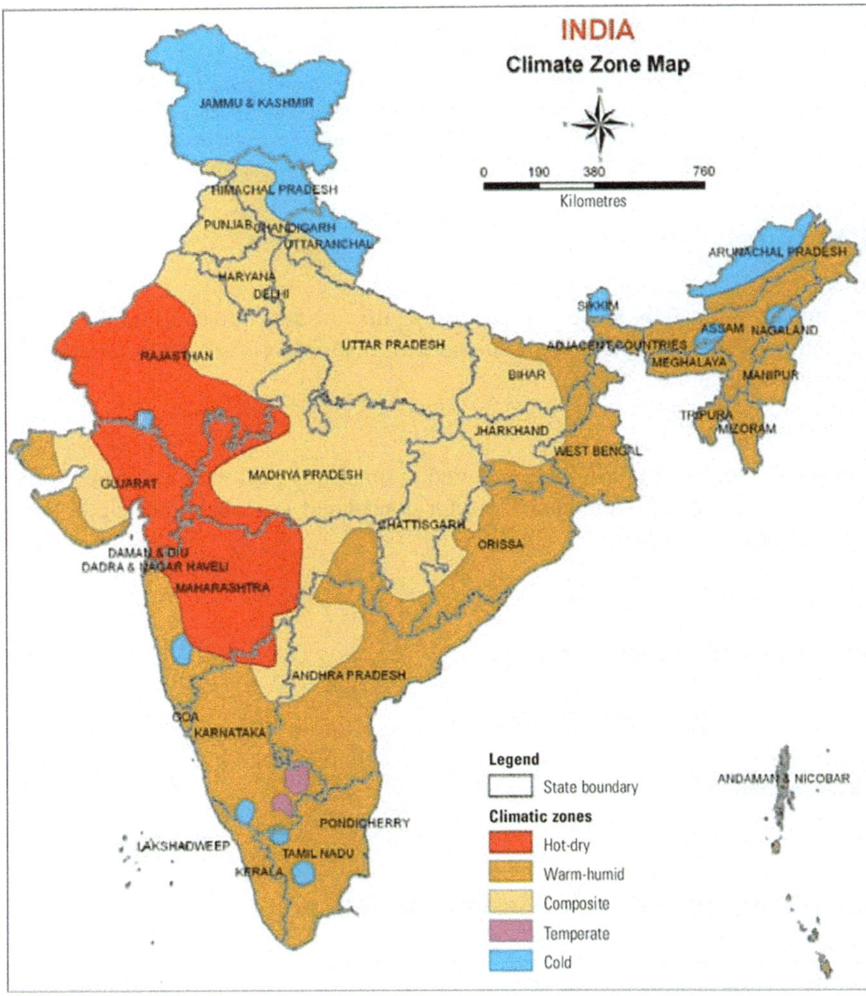

Fig. 6.1 Climate map of India [3]

Fig. 6.2 External view of the Pocharam campus [4]

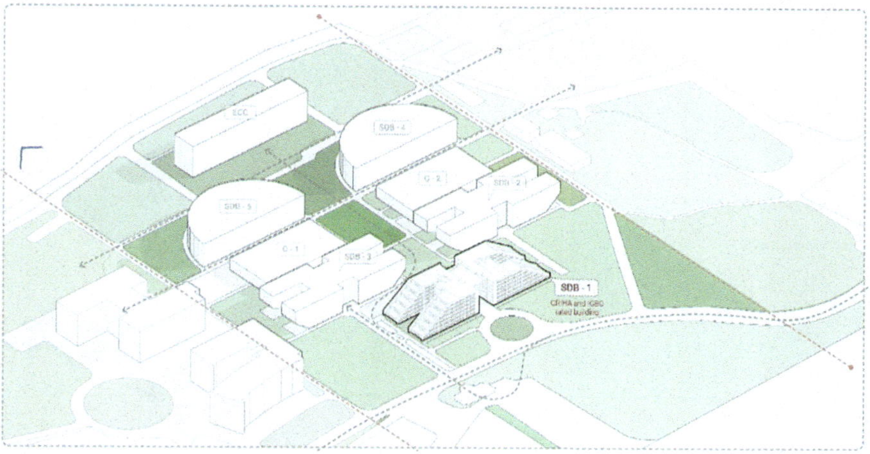

Fig. 6.3 Location of the green-rated building on the office campus

6.5.1 Climatic Analysis

Hyderabad has a composite climate, with a rainy season from June to October, due to the monsoon, and a dry season from November to May. The city is in south-central India, at 17° north latitude and 550 m (1800 ft) above sea level. The environment variables are also graphically indicated in Fig. 6.4.

- Summer lasts from March to June, with temperatures ranging from 25 to 40 °C. The city can experience heat waves.
- The monsoon season lasts from July to September, with heavy rainfall and thunderstorms. While temperature shows comfort during this period, humidity exceeds comfort levels.
- Winter in Hyderabad lasts from October to February, with temperatures ranging from 10 to 25 °C, and is just within the comfort range, sometimes dipping below.
- The temperature remains largely comfortable. It peaks in the summer months and tends to dip in the nights of the winter months.
- Humidity is on the higher side in monsoon months and varies in summer, indicating somewhat dry summers.
- Radiation is high in the summer months of March to June. From June onwards diffused radiation is higher indicating cloud cover.
- Cool winds from the west optimize in the monsoon period. Winds blow from East and South-East in Winter and Summer periods.

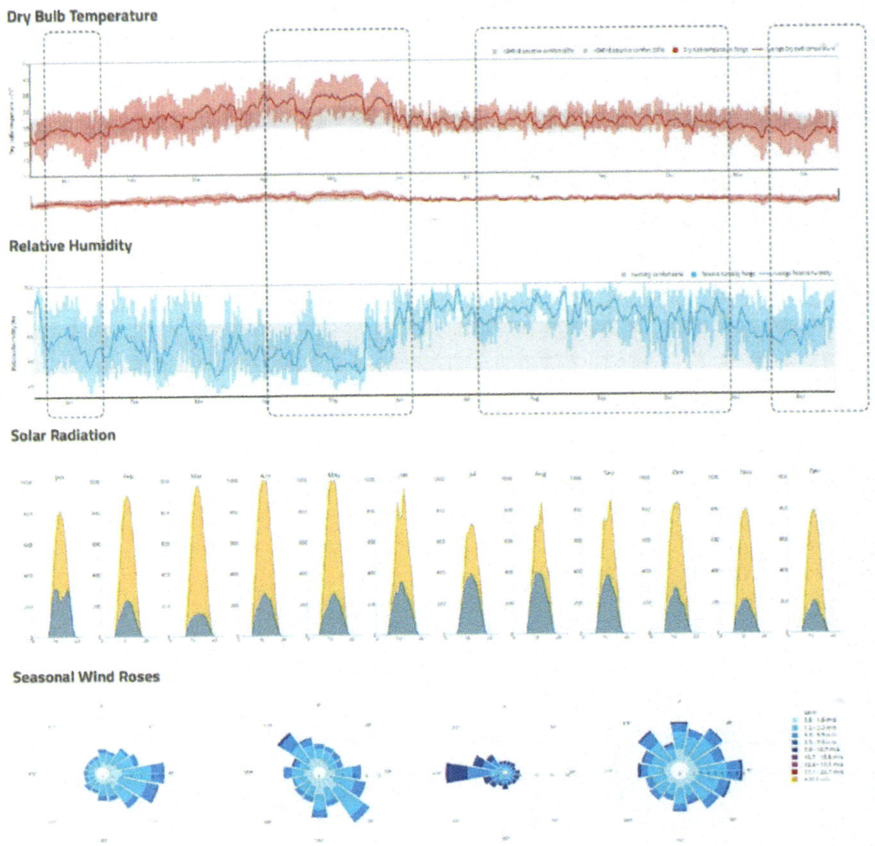

Fig. 6.4 Temperature, humidity, radiation, and wind variables plotted using tool [5]

Passive strategies used:

- The building's longer axis is oriented in the East–West axis in order to reduce solar heat gain.
- 78% of living areas are day-lit and window to wall ratio is restricted to <38% to reduce solar heat gain inside the building.
- The building is designed conducive to natural ventilation.

As discussed in Chap. 4, materials and structural systems are given more focus, in line with the guidelines developed.

6.5.2 *Details of Base Case Building*

The base case building is designed on the basis of the design and planning criteria of the Infosys, Pocharam Campus building. The base case building has the following features:

- The building is a 6-storey office building with a capacity of around 2400 people.
- It has open and closed offices along with common spaces, lobbies, meeting rooms, etc.
- It is an RCC-framed structure with two typical wings.
- One wing is conditioned using a VAV system and the other wing is conditioned using the radiant cooling system to demonstrate energy savings.

The building has a total height of 35 m. The width of the floor plate is limited to 16 m to ensure adequate daylight from north and south-facing windows (Figs. 6.5, 6.6 and 6.7).

Design-based passive strategies are elaborated, with respect to the context of Hyderabad (Fig. 6.6).

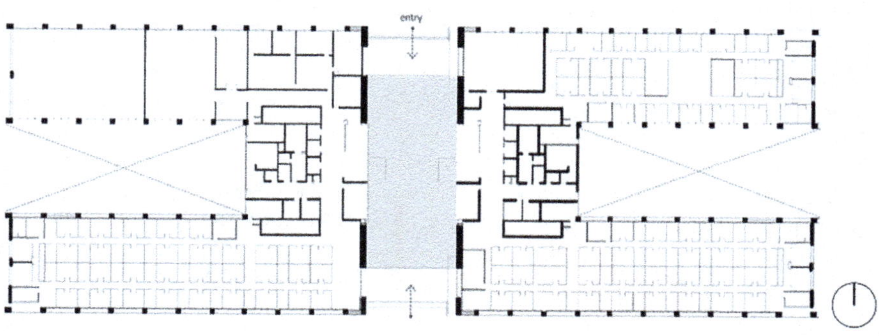

Fig. 6.5 Ground floor plan. Source: Author

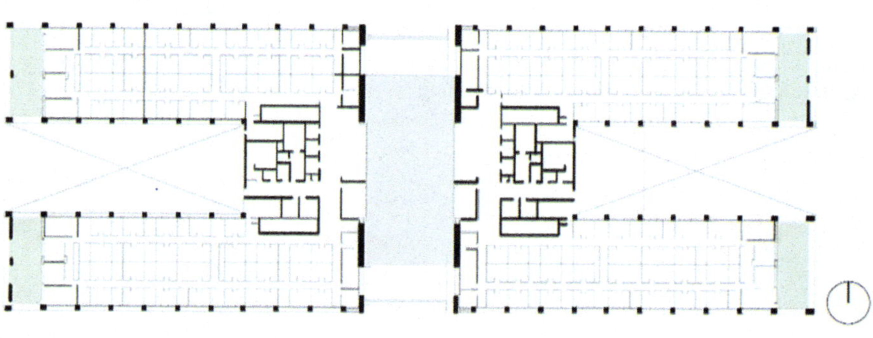

Fig. 6.6 First floor plan. Source: Author

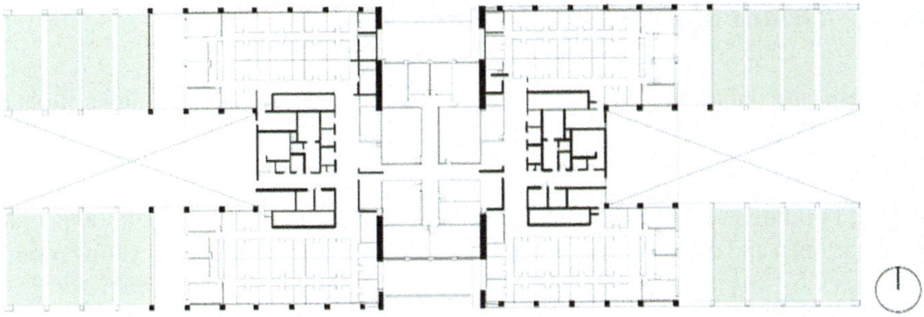

Fig. 6.7 Fourth floor plan. Source: Author

The radiation on the surface of the building is evaluated as seen in Fig. 6.8 and broken down in Fig. 6.9 and could be used in passive design iterations. For any iterations, it would be beneficial to ensure that any change in form, orientation, windows, and shading devices, does not exceed the surface area or incident radiation of the base case.

Exposed surface area

North façade—3785 m^2
East facade—2180 m^2
West facade—2180 m^2

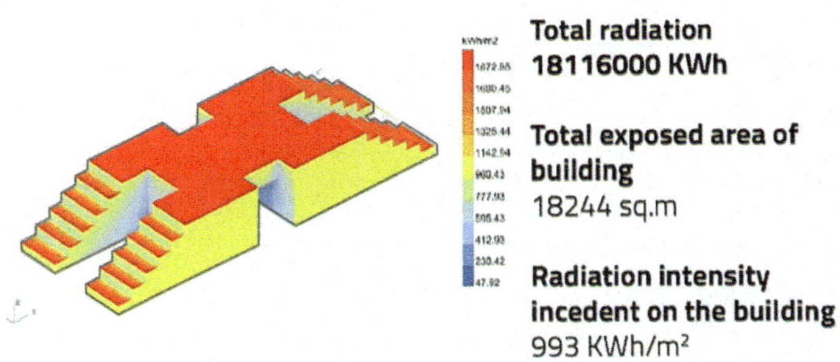

Fig. 6.8 Radiation study of the base case building

Fig. 6.9 Radiation on surfaces of walls and roofs

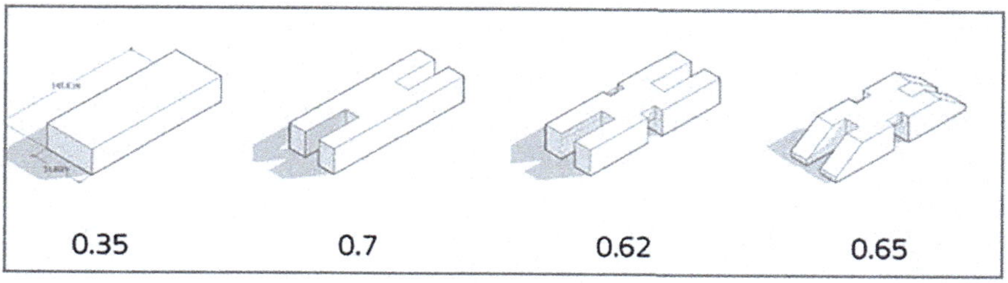

Fig. 6.10 Aspect ratio of buildings

South facade—3785 m^2
Exposed surface area
Roof area—3589 m^2
Terrace area—2175 m^2
Total radiation incident on the roof
Roof—6,721,500 KWh
Terraces—3,559,500 KWh
Total—102,81,000 KWh

Total radiation incident on the walls

North—1,339,900 KWh
South—3,466,600 KW.
East—1,757,600 KWh
West—1,477,200 KWh
Total—80,41,300 KWh

Though many elements of the proposed base case are discussed such as form and orientation from Fig. 6.10, in line with the guidelines developed in Chap. 4, the comparison of iterations for carbon reductions largely revolves around materials, systems, and structure. This is due to the fact that passive design strategies respond to context and influence both embodied and operational emissions.

6.5.3 Baseline Case

Slab

Re-enforced concrete (M25) with 40% PFA replacement
Indian Construction Materials Database

Columns and beams

Re-enforced concrete (M25) with 30% PFA replacement
Indian Construction Materials Database

Walls

N–W: Clay insulated tiles over double block
E–W: Double brick cavity walls
Custom Indian Materials Database

Windows and shading

Aluminum framed double glazed windows with argon gas and air cavity
Aluminum shades and louvers for horizontal Shading devices
Frosted glass fins for vertical shading devices
Custom Indian Materials Database
The façade and materials are seen in Figs. 6.11, 6.12 and 6.13.
Whole building elements were studied.

1. Structure
2. Wall materials
3. Window
4. Shading
5. Cladding materials.

Figure 6.12 indicates that RCC slabs, columns, and beams form the major share of materials in the building, followed by envelope masonry material. The remaining materials form <10% of the building volume.

Only civil materials are considered in the study. Interiors and MEP system materials are considered negligible and excluded.

For an area of 26,000 m^2, the embodied carbon per m^2 of the building is 451.96 $kgCO_2e/m^2$.

Assuming 60 years of the life of the building, the total embodied carbon value is 7.53 $kgCO_2e/m^2$/year.

The materials with the highest CO2e value per m3 are (1) Concrete elements of the building (2) Insulation material (EPS), Aluminum louvers and frames, followed by (3) Steel rebar, and then the (4) DGU of glass.

6.5.3.1 Energy Analysis—Inputs and Baseline

Zones are considered for the energy model as seen in Fig. 6.14.

- Open office.
- Closed office.
- Toilets.
- Lift core.
- Staircase core.
- Lobby.

Inputs considered:
Occupancy and use—9 AM to 5 PM.

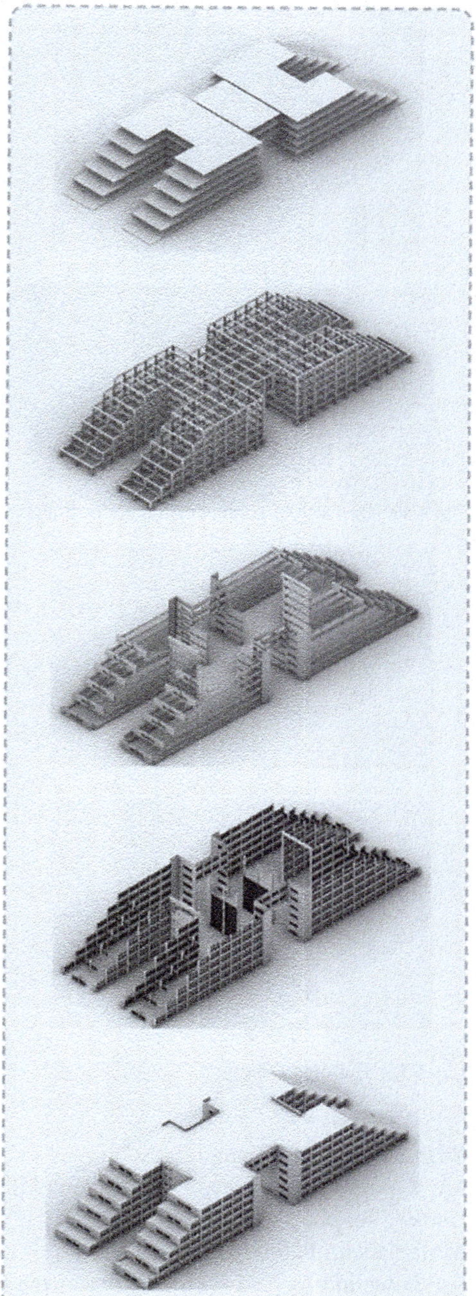

Fig. 6.11 Materials exploded from the building. Source: Author

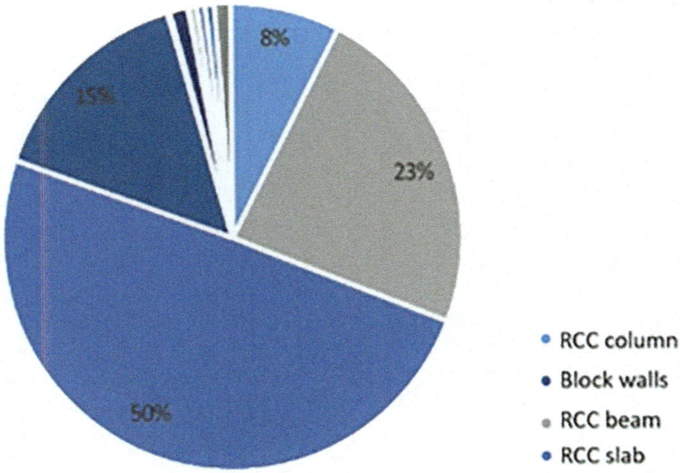

Fig. 6.12 Breakup of volume of materials. Source: Author

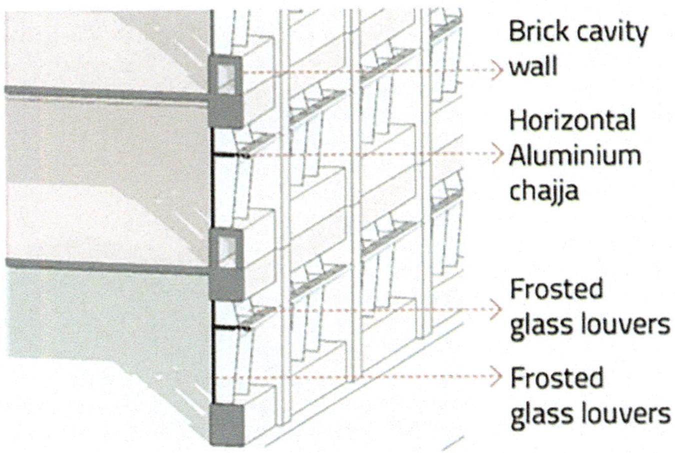

Fig. 6.13 Building section

People load—2400 people.
Air conditioning system—VAV system chiller with air pump.
Cooling set point—26°.
Materials and constructions:
As shown in Fig. 6.11 with columns as internal mass, form, orientation, WWR, and shading devices as seen in Fig. 6.15.
The life of the building base case from Fig. 6.15 is considered to be 60 years and the associated results for carbon emissions are as follows (Table 6.1):

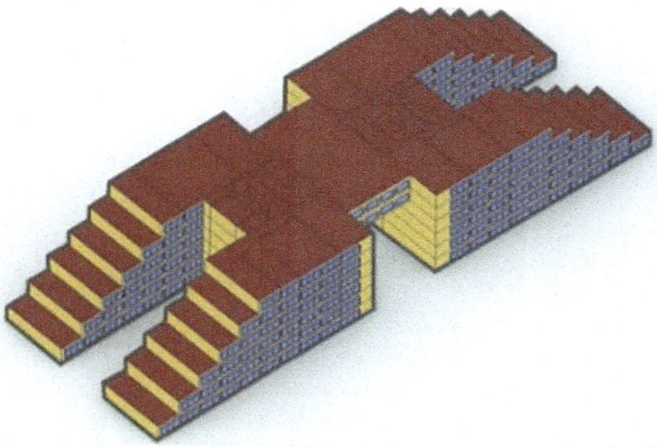

Fig. 6.14 Energy model of the base case of the building

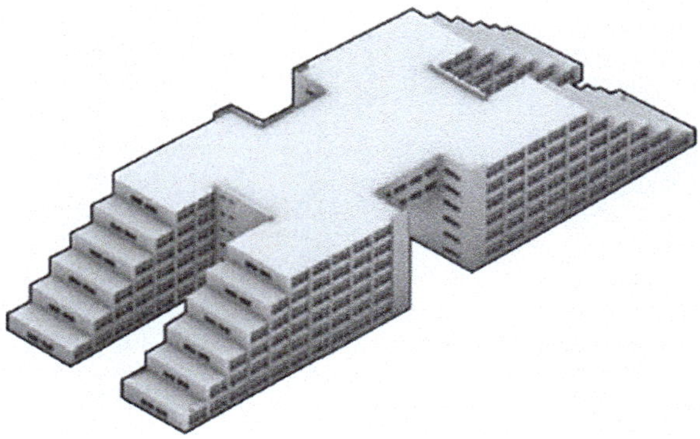

Fig. 6.15 3D model of the baseline scenario

Table 6.1 Resultant EUI of the energy study

End energy use	**EUI** (KWh/m^2/yr)
Cooling	33.3
Interior lighting	18.4
Electrical equipment	27.8
VAV system fans	2.3
Pumps	6.7
Heat rejection	1.8

Total OC per year—73.155 KgCO_2e/m^2/year
Total EC—421.6 KgCO_2e/m^2 or 7.03 KgCO_2e/m^2/year

The graph in Fig. 6.16 shows the contribution of materials to embodied carbon and the carbon balance of the building over the life cycle, respectively.

It indicates that concrete is a major contributor followed by aluminum and steel for embodied carbon emissions. So, these elements can be looked at iteratively.

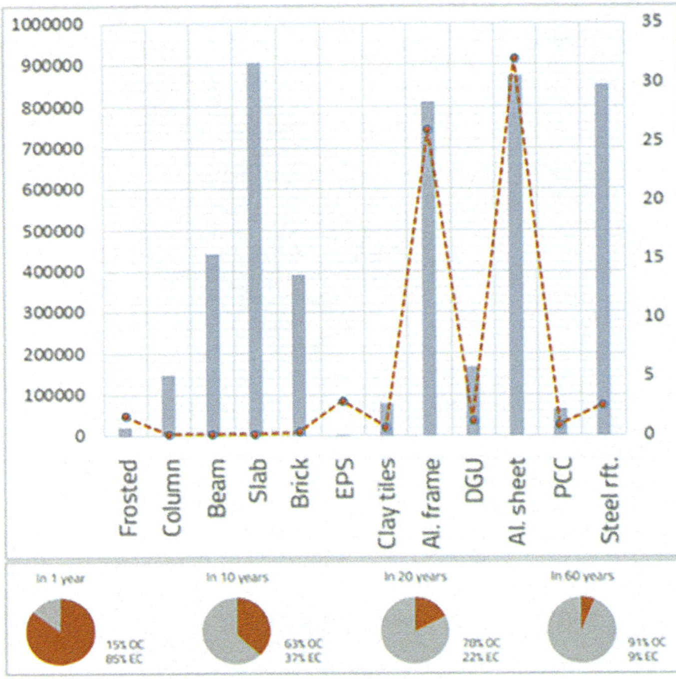

Fig. 6.16 Carbon emissions of the base case

The base case considers one-half of the building and iterates on this base using the above guidelines and evaluates the impact on operational energy use and carbon along with the change in embodied carbon values.

6.6 Summary

LCA studies are described by their complexity. Unlike energy metrics like EUI, a carbon-based LCA metric is associated with upfront emissions, use, and the emissions associated with disposal as well. The share of these emissions in a building can lead to complex, project, time, and context-specific conclusions. This is essential to understand the complete carbon emissions or the complete environmental profile of a building in a particular context.

However, key challenges particularly in India is the availability of data and tools. Most Indian LCAs conducted in tools do not accommodate Indian materials and products. The government of India or any national-level entity has yet to initiate an India-specific database. This poses a major challenge for LCA studies in India.

Due to this challenge, there may be a lack of accuracy associated with detailed LCA. However, for comparative studies or design decisions at the early stage of a building, it is possible to do LCA studies with a clear definition of the goal, scope, system boundary, and functional unit studied.

As such, there is no requirement to wait for more clarity in LCA data. It is imperative to start addressing wholistic emissions using LCA in buildings. The methodology of LCA can be applied immediately to buildings with the available tools and data. The presence of these challenges and limitations should, however, be noted during the LCA.

The LCA methodology adopted in this book is discussed with a case study of an office building. It is a concrete framed 6-storey building with an estimated life of 60 years considered for the study. The operational and embodied carbon emissions are 73.15 $kgCO_2e/m^2/year$ and 7.03 $kgCO_2e/m^2/year$. This indicates the total impact of embodied carbon as 10% compared to operational carbon emissions across the life of the building. Though this may seem negligible, it should be noted that it cannot be modified after construction. Operational emissions can be reduced using decarbonization measures, but most of the embodied carbon emissions are released before the building is built. This means it needs to be addressed immediately to meet future decarbonization and net-zero targets. This building carbon analysis will form the baseline of the comparative carbon emissions review. The objective is to study the impact of the guidelines from Chap. 2, and the possible reduction in each case compared to this baseline model. A review of these impacts and guidelines with the green building rating systems can give some direction to the future development of green certifications and green buildings.

References

1. Tangible (2023) A call to action for the real estate industry: it's time to address embodied carbon. Retrieved from Linkedin. https://www.linkedin.com/pulse/call-action-real-estate-industry-its-time-address-embodied/
2. Hussain M (2022). Linked in. Retrieved from Growth of Hyderabad as a business hub: Is it over? https://www.linkedin.com/pulse/growth-hyderabad-business-hub-over-misbah-hussain/
3. Bureau of Indian Standards (2005) National building code of India 2005. Bureau of Indian Standards
4. The Hans India (2015) Infosys Hyderabad facility to be green. Retrieved from The Hans India. https://www.thehansindia.com/posts/index/Telangana/2015-12-29/Infosys-Hyderabad-facility-to-be-green/196207?infinitescroll=1
5. Betti G, Tartarini F, Nguyen C, Schiavon S (2023) CBE Clima tool: a free and open-source web application for climate analysis tailored to sustainable building design. Retrieved from Build. Simul. https://doi.org/10.1007/s12273-023-1090-5

Chapter 7
Structural Material Iterations

7.1 Introduction

While operational emissions from buildings in use have been the primary concern in the recent past, the industry is now shifting toward embodied carbon. Embodied carbon has an immediate impact on the environment, through building materials—and the emissions released during extraction, production, and construction phases.

Of these, structure and enclosure materials account for the largest portion of embodied carbon emissions, as discussed in Chap. 4, almost at 50%. They also represent the greatest opportunity for reducing a building's overall embodied carbon footprint. This section describes how the guidelines developed in Chap. 4, for structural materials, envelope materials, and systems are applied to the baseline case of the building.

The quantification of these iterations highlights the impact of the iterations and is used to discuss feasible strategies for carbon-conscious design and decarbonization of buildings.

How are the guidelines applied:

Parameters: This describes the basic principles summarized above for low-carbon strategy implementation.
Guideline: Possible guidelines that can be developed based on the parameters.
Building typology: This describes the typology of the building where the guideline can be applied. This may include new buildings or retrofit/refurbishment of existing buildings.
Stage of design: It indicates the stage in the life of a building, where these parameters can have more impact. This can include concept stage, design development, detailed design, construction, use, and end of life of the building. Most of the guidelines are meant to target the early stage of building design.

S. Mohan and Vijayalaxmi J., *Embodied and Operational Carbon in Buildings*,
SpringerBriefs in Architectural Design and Technology,
https://doi.org/10.1007/978-981-97-7187-5_7

Key considerations: There are other factors that could influence decision-making in relation to the guideline discussed. This is parameter-specific and could include key variables like structural stability, material availability, mix specifications, etc.
Application: For this exercise, the applications are a narrowing of certain best practices and strategies found in case studies and alternative research. These applications are numerous and will vary with project and context-specific factors (Table 7.1).

As discussed in Chap. 4, the building design guidelines broadly deal with four categories.

The alternatives and pathways considered are illustrated in Fig. 7.1.

For passive design:

- The structural material and system alternatives are targeted at concrete and steel reduction. Agro-waste, industrial waste, and natural materials are partially substituted for cement to reduce the volume and thereby the carbon emissions of the material.

  The structural system alternatives use recycled steel, bamboo, and other methods of construction to reduce or compensate for the production emissions of reinforced concrete.
- Alternative materials which are bio-based, natural, or waste based have been substituted for masonry as well. These criteria are prevalent among current rating systems as well and comparatively quantifying the impact may aid target setting.

The alternatives illustrated in Fig. 7.1 are not comprehensive. There may be various materials and types of systems available in a given local context. The most accessible source to reduce transport emissions should be considered. This theoretical model, however, assumes a series of 7 iterations for the purpose of the structural impact study and derivation of comparative inferences. The following sections describe the boundaries, assumptions, and results of the guidelines.

These alternate materials are studied in comparison with the baseline discussed in Chap. 6. All references to the baseline building, therefore, refer to the following values.

Total OC per year—73.155 $KgCO_2e/m^2/year$
Total EC—421.6 $KgCO_2e/m^2$ or 7.03 $KgCO_2e/m^2/year$.

Table 7.1 Methodology to apply the design guidelines with respect to context

S.no.	Parameters	Guideline	Building type	Stage	Considerations	Application
1	Type of structural system	Structural system types to be considered during design	New buildings	Conceptual design	Available local material and structural soundness are key considerations	Steel-based structural system (with maximum recyclability)
						Concrete framed structure
2	Structural material used	Percentage of the cement in concrete buildings to be replaced with bio-based	New buildings	Concept design and Design development	Locally produced agricultural and industrial waste which can be used as cement replacement in green concrete, without compromising structural properties	Agro-based green concrete mixes (RHA, BLA, etc.)
						Use of precast concrete for the structure
						Industrial waste-based concrete (Fly ash, GGBS, etc.)
						Use of bamboo in concrete slabs

(continued)

Table 7.1 (continued)

S.no.	Parameters	Guideline	Building type	Stage	Considerations	Application
3	Alternative materials used	Use of responsibly sourced natural materials like strawbales, cellulose insulation, rammed earth, CSEB, etc., and waste-based materials like Fly ash, PFA, GGBS, etc	New buildings/ retrofit existing buildings	Concept design and design development	Key considerations include specification and mixes of materials and volume of material used/ replaced in the building	Use of bio-based materials (Porotherm blocks, etc.)
						Use of natural materials (Rammed earth, CSEB, etc.)
					Locally produced agricultural and industrial waste which can be used as cement replacement in green concrete for envelope materials	Use of industrial waste, and salvaged materials (FAlg, AAC blocks, etc.)
						Use of roof alternatives (Green roof, filler slab, etc.)
4	Passive design	Use of orientation, WWR, shading devices, etc	New buildings	Conceptual design	Context, climate, geographic and cultural area, etc. are key considerations	The application in this book is limited, as the study happens earlier in the design process, is specific to the context, and can have a range of iterations

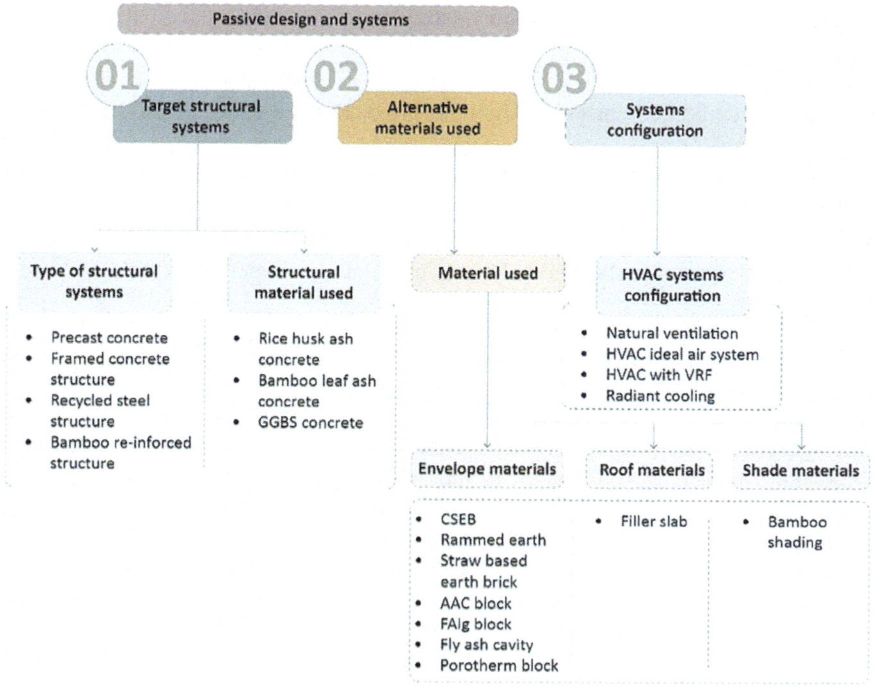

Fig. 7.1 Design-based iterations considered

7.2 Target Structural Systems

7.2.1 Precast Concrete

The type of structural system influences embodied carbon, particularly in its material phase. Due to the need for faster construction, Precast concrete is a very popular method of construction and is establishing itself as a key construction method for the future.

Key considerations:

- The columns and beams of the baseline building are considered to be precast elements.
- Wall panels and slabs are not iterated here and are kept as a base case material, which is in-situ concrete.

Embodied and Operational Carbon Analysis

For embodied carbon of the building with precast concrete:

- The total embodied carbon of the building calculated using precast elements is 5051104 kgCO_2e.
- These embodied carbon emissions normalized over the area of the building is 453 kgCO_2e/m^2.
- When this normalized embodied carbon is divided by the lifespan of the building (60 years) **annual embodied carbon** is calculated, which is 7.6 kgCO_2e/m^2/year.

The embodied carbon is represented in Fig. 7.2 with the red lines of the column and beam elements indicating the reduction from the baseline case.

For operational carbon of the building with precast concrete:

The operational carbon is calculated using Rhino and Grasshopper software with the Open Studio engine. Changes in thermal properties based on the material are made for precast concrete material with the same energy model properties for the remaining parameters, as the baseline case.

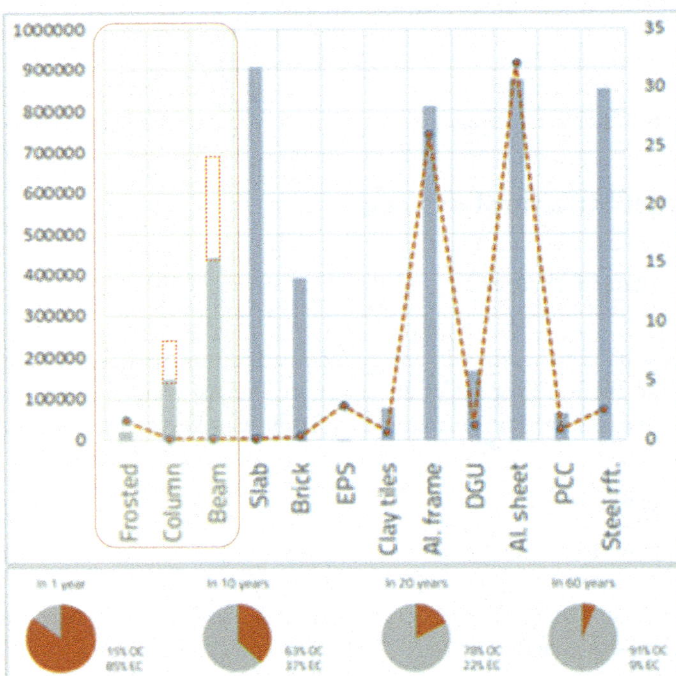

Fig. 7.2 Embodied carbon of elements using precast concrete over the base case and balance of carbon emissions over the life of the building

The resulting EUI is then used to calculate Operational carbon using India's electricity carbon emission factor. This method is followed for all the iterations with material changes to compare operational carbon.

- The EUI of the proposed building is 90.316 $KWh/m^2/year$.
- The **operational carbon** using India's CE factor is 73.15 $kgCO_2e/m^2/year$.

The graph of the carbon emissions from concrete beams and columns increased as precast concrete is more energy intensive between A1 and A3 of the lifecycle.

Inferences

The beam and column carbon factors have increased as precast requires more energy to manufacture, but less time to install.

Further, against the baseline case, the embodied carbon shows a 6% increase, while operational carbon remains the same.

With a precast system, the construction, and installation (A5) part of the lifecycle, may give larger carbon savings, than the material and manufacturing.

7.2.2 *Rice Husk Ash Concrete*

Agro-waste also tends to have low embodied carbon and thermal properties due to its bio-based nature. The second iteration replaces concrete with Rice husk ash-based agro-concrete.

The biggest agricultural waste generated in India is rice husk [1].

Key considerations:

- The ratio of replacement for rice husk ash is considered as 15% of the cement volume for structural concrete taken from the source in Table 7.2.
- The structural concrete is replaced only in the column and beam structure of the baseline building.
- In India, RHA is produced by burning rice husk in the open air, resulting in environmental pollution, which should also be considered and mitigated while using RHA.

Embodied and Operational Carbon Data

For embodied carbon of the building with rice husk ash concrete:

- The total embodied carbon of the building calculated using the RHA concrete is 4759313 $kgCO_2e$.

Table 7.2 Mix of Rice Husk Ash (RHA) used and the source

S. no.	Material used	EC	Sources
1	Rice husk ash	**423.4**	[1, 2]

- These carbon emissions are normalized over the building area, and this results in **426.9 $kgCO_2e/m^2$**.
- When these normalized embodied carbon emissions are divided by the lifespan of the building (60 years) **annual embodied carbon** is calculated, which is **7.12 $kgCO_2e/m^2/year$**.

The embodied carbon is represented in Fig. 7.2 with the red lines of the column and beam elements indicating the reduction from the baseline case (Fig. 7.3).

For operational carbon of the building with rice husk ash concrete:

As discussed in the Precast concrete iteration, The operational carbon is calculated using Rhino and Grasshopper software with the Open Studio engine. Changes in thermal properties based on the material are made for Rice husk ash based on cited sources. The remaining parameters of the energy model remain the same as the baseline case.

- The EUI of the proposed building is 90.316 $KWh/m^2/year$.
- The **operational carbon** using India's CE factor is 73.15 $kgCO_2e/m^2/year$.

The replacement factor of RHA in cement is lesser than the fly ash replacement in concrete, which is used in the base case. Therefore, the total embodied carbon does

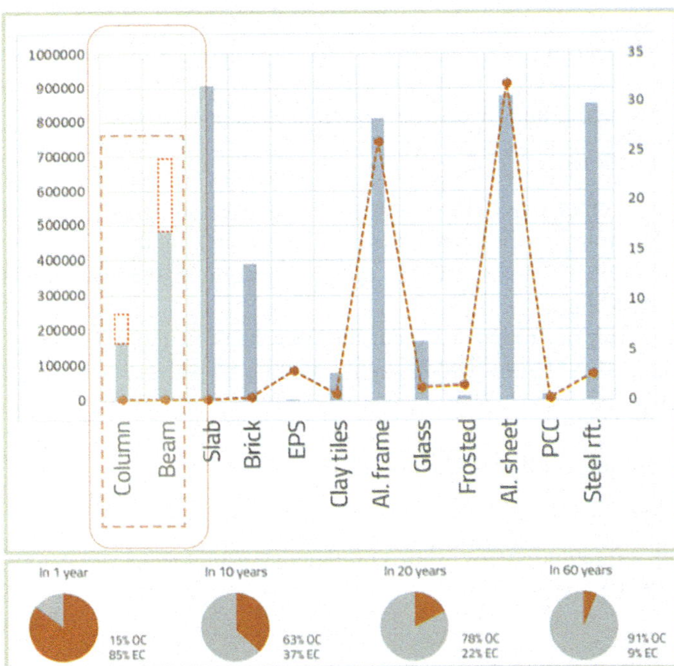

Fig. 7.3 Embodied carbon of elements using Rice Husk Ash (RHA) concrete over the base case and balanced carbon emissions over the life of the building

not show a reduction from the base case. The greater the percentage of replacement, without compromising structure, the larger the carbon saving will be.

Inferences

Rice husk ash concrete at 15% replacement of volume of cement gives only 2% embodied carbon raise over fly ash of 30% volume. If structurally compatible, more quantity of RHA be used in concrete to reduce carbon footprint.

7.2.3 Bamboo Leaf Ash Concrete

Similar to RHA, agrowaste also tends to have low embodied carbon and thermal properties due to its nature.

Bamboo is widely grown and used for various purposes in India, resulting in large amounts of solid waste, and this can be used as a replacement for cement in concrete (Table 7.3).

Key considerations:

- The ratio of replacement is considered as 10% of the cement volume for columns and beams (Table 7.3).

Embodied and Operational Carbon Data

For embodied carbon of the building with bamboo leaf ash concrete:

- The total embodied carbon of the building using bamboo leaf ash concrete is 4748267 $kgCO_2e$.
- These carbon emissions normalized per building area **425.8 $kgCO_2e/m^2$**.
- These normalized embodied carbons are divided by the lifespan of the building (60 years) to calculate **annual embodied carbon**, which is **7.09 $kgCO_2e/m^2$/year**.

The embodied carbon is represented in Fig. 7.4 with the red lines of the column and beam elements indicating the reduction from the baseline case.

For operational carbon of the building with bamboo leaf ash concrete:

The same methodology, discussed in the above sections using Rhino and Grasshopper software with Open Studio engine is used for energy analysis and to arrive at operational carbon emissions of the proposed iteration.

- The EUI of the proposed building is 90.316 KWh/m^2/year.

Table 7.3 Mix of Bamboo Leaf Ash (BLA) used and the source

S. no.	Material used	EC	Source
1	Bamboo leaf ash	**371.02**	[3]

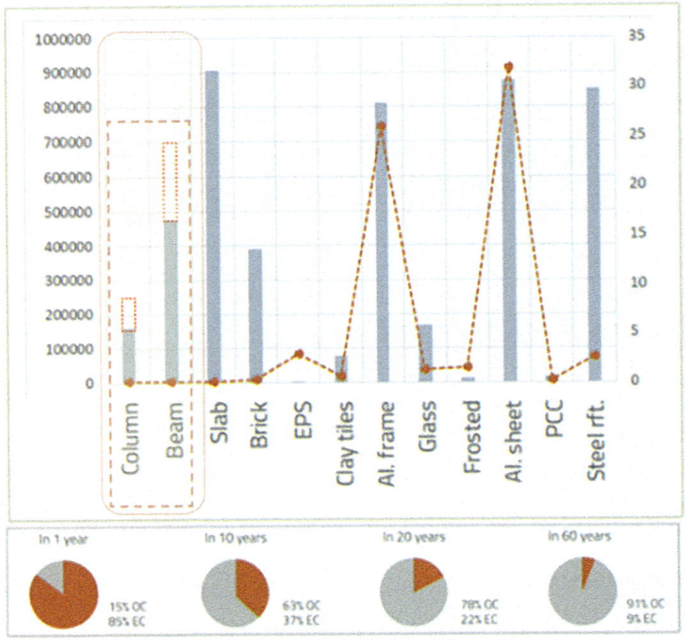

Fig. 7.4 Embodied carbon of elements using Bamboo Leaf Ash (BLA) concrete over the base case and balance of carbon emissions over the life of the building

- The operational carbon using India's CE factor is **73.15 $kgCO_2e/m^2/year$**.

The replacement factor of BLA in cement is lesser than the fly ash replacement in concrete, which is used in the base case. Yet, the total embodied carbon shows a reduction from the base case by 2%. The greater the percentage of replacement, without compromising structure, the larger the carbon saving will be.

Inferences

Bamboo leaf ash concrete at 10% replacement of volume of cement gives around 2% embodied carbon savings over fly ash of 30% volume. If structurally compatible, more quantities of BLA be used in concrete to reduce carbon footprint.

7.2.4 GGBS Concrete

The fourth iteration used is the replacement of GGBS in concrete from the baseline. GGBS is industrial waste and can be used to replace concrete up to 75–80%, according to US and British standards. However, cement manufacturers approximately use 50% replacement of cement in concrete.

Table 7.4 Mix of GGBS used in the concrete

S. no.	Material used	EC	Sources
1	Ground granulated blast slag	**376.09**	[1, 4]

Key Considerations:

- Ratio of replacement is considered as 40% of the cement volume for columns, slabs, and beams as indicated in the Table 7.4.
- Concrete beams, columns, and slabs, contain GGBS in the same percentage of fly ash in the base case Table 7.4.

Embodied and Operational Carbon Data

For embodied carbon of the building with GGBS concrete:

- The total embodied carbon of the building using GGBS concrete for the structural elements is 4672357 $kgCO_2e$.
- These emissions are normalized over the building area as **419 $kgCO_2e/m^2$**.
- The normalized embodied carbon is divided by the lifespan of the building (60 years) to calculate **annual embodied carbon,** which is **6.98 $kgCO_2e/m^2/$year**.

The embodied carbon is represented in Fig. 7.5 with the red lines of the column and beam elements indicating the reduction from the baseline case.

For operational carbon of the building with GGBS concrete:

The same methodology, discussed in the above sections of using Rhino and Grasshopper software with Open Studio engine is used for energy analysis and to arrive at operational carbon emissions of the proposed iteration.

- The EUI of the proposed building is 90.316 KWh/m^2/year.
- The **operational carbon** using India's CE factor is **73.15 $kgCO_2e/m^2$/year**.

Inferences

GGBS concrete at 40% replacement of volume of cement gives around 2% embodied carbon savings over fly ash of 40% volume. If structurally compatible, more quantity of GGBS be used in concrete to reduce carbon footprint.

7.3 Summary

There is very little variation in operational carbon, for changes in **concrete mixes**.

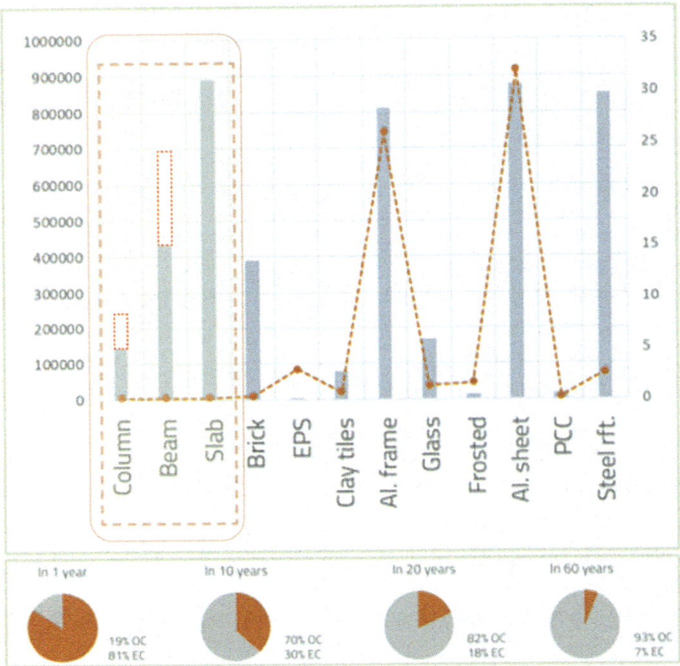

Fig. 7.5 Embodied carbon of elements using GGBS concrete over the base case and balance of carbon emissions over the life of the building

However, the embodied carbon varies with every **significant variation in the concrete mixes**, showing **Fly Ash and GGBS concrete as options with the least embodied carbon emissions,** as shown here.

The embodied carbon of the materials shown in the study is **per m² of the building.**

The embodied carbon using conventional **OPC-based concrete is calculated to be 448 kgCO₂e/m² of the building.**

The amount and elements of replacement of cement in concrete are also shown in Fig. 7.6.

Both GRIHA and IGBC show percentage reductions from a conventional baseline.

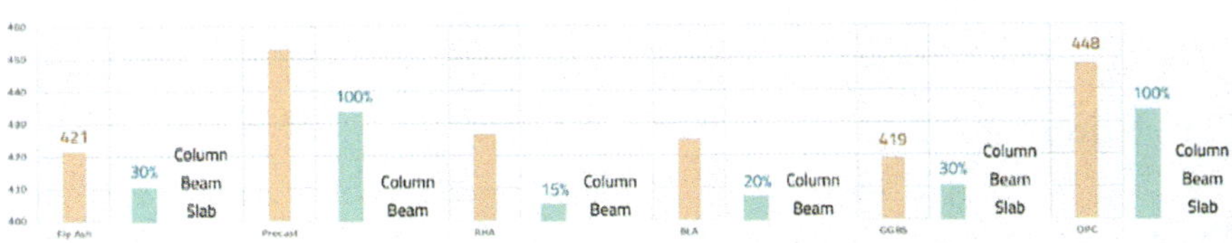

Fig. 7.6 Charts for structural iterations of the building

Table 7.5 Results based on changes in structural material

Parameter	Embodied carbon ($kgCO_2e/m^2$)	Operational carbon ($kgCO_2e/m^2/yr$)	EC change compared to a conventional case (%)
Base case	421.6	73.15	−11
Conventional case	476	81	
Precast concrete	453	73.15	−5
Rice husk ash concrete	426	73.15	−11
Bamboo leaf ash concrete	425	73.15	−11
GGBS concrete	419	73.15	−12

Here, an already reduced concrete mix with 40% Fly Ash, has been considered as the baseline, so the reduction percentages are low.

In case they are compared to a conventional building, then the percentage reduction is around 7–8%.

This is described below in a Table 7.5.

7.4 Type of Structural System Changes

7.4.1 Bamboo Reinforced Structure

Bamboo is widely grown and used for various purposes in India. It is highly flexible, lightweight, and economically viable. Compared to steel, bamboo shows great tensile strength and is a very strong material [5]. Bamboo is used as an option to understand the impact against the conventional base case.

Key considerations:

- Like steel reinforcement, bamboo-reinforced concrete follows the same construction techniques, mix proportions, and designs.
- Concrete beam, column, and slab elements have bamboo reinforcement iterations over the base case of steel reinforcement.
- The species, water absorption, and durability of the bamboo a key considerations to be considered as reinforcement for concrete.

Embodied and Operational Carbon Data

For embodied carbon of the building with bamboo reinforcement:

- The total embodied carbon of the building using bamboo reinforcement and concrete from the base case is 3904147 $kgCO_2e$.
- The carbon emissions are normalized per building area **350.14** $\mathbf{kgCO_2e/m^2}$.

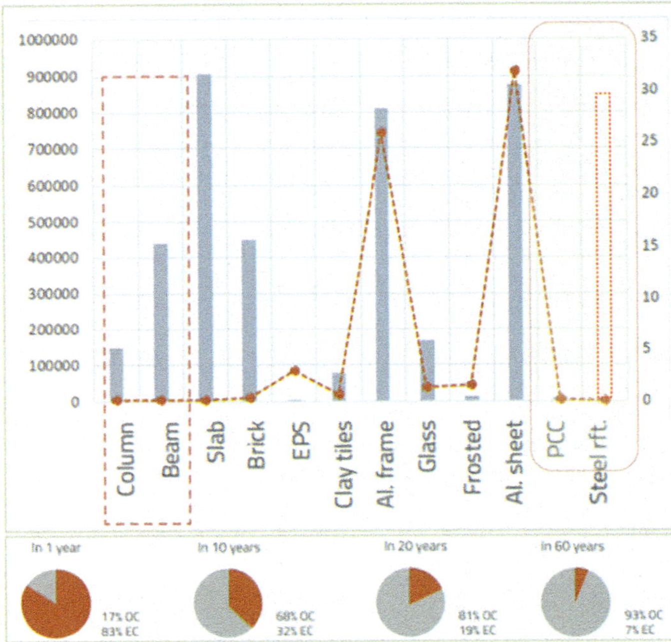

Fig. 7.7 Embodied carbon of elements using Bamboo reinforced concrete over the base case and balance of carbon emissions over the life of the building

- This normalized embodied carbon is divided by the lifespan of the building (60 years) to calculate **annual embodied carbon**, which is **5.83 kgCO_2e/m^2/year**.

The embodied carbon is represented in Fig. 7.7 with the red lines of the column and beam elements indicating the reduction from the baseline case.

For operational carbon of the building with bamboo reinforcement:

The same methodology, discussed in the above sections of using Rhino and Grasshopper software with Open Studio engine is used for energy analysis and to arrive at operational carbon emissions of the proposed iteration.

- The EUI of the proposed building is 90.316 KWh/m^2/year.
- The **operational carbon** using India's CE factor is **73.15 kgCO_2e/m^2/year**.

Steel reinforcement is a major contribution to the carbon emissions of the base case.

Key consideration includes the durability of bamboo which is lesser compared to steel and the low modulus of elasticity. It also has limitations on application and is largely in the R&D phase, as a reinforcement alternative.

Inferences

On a comparative scale, the applicability of bamboo reinforced concrete is not as straightforward as the replacement of cement in concrete mixes. However, on a theoretical basis, high-carbon emission reductions can be achieved by reducing or replacing the carbon emissions of steel. In this iteration, almost 17% of the embodied carbon emissions can be reduced.

7.4.2 Recycled Steel Structure

The steel content in buildings can also be addressed, by increasing the recycled content of steel. Currently, the recycling rate of steel in India is way below the global market of 85%. India recycles only 22% of the steel produced [6]. The National Steel Policy 2017 and Steel Scrap Recycling Policy 2019, outline the importance of using ferrous scrap to reduce the GHG emission intensity of the steel sector. However, it does not provide a clear roadmap or incentives to increase the usage of scrap and organize scrap collection in India [6].

Key considerations:

- Since the recycling rate in India is lower than the global average, the current iteration considers a recycling rate of 20% compared to the global average of 83% for the recycling rate of steel, which is used in the base case.
- The ICE database value for the global average of recycled steel is 0.45 kgCO_2e/kg compared to the 2.6 kgCO_2e/kg in India.

Embodied and Operational Carbon Data

For embodied carbon of the building with recycled steel:

- The total embodied carbon of the building using 20% recycled steel is 7368185 kgCO_2e.
- These emissions are normalized per building area of **660 kgCO_2e/m^2**.
- The normalized embodied carbon is divided by the lifespan of the building (60 years) to calculate **annual embodied carbon**, which is **11.01 kgCO_2e/m^2/year**.

The embodied carbon is represented in Fig. 7.8 with the red lines of the column and beam elements indicating the reduction from the baseline case.

For Operational Carbon of the Building with Recycled Steel:

The same methodology, discussed in the above sections using Rhino and Grasshopper software with Open Studio engine is used for energy analysis and to arrive at operational carbon emissions of the proposed iteration.

- The EUI of the proposed building is 90.5 KWh/m^2/year.
- The **operational carbon** using India's CE factor is **73.5 kgCO_2e/m^2/year**.

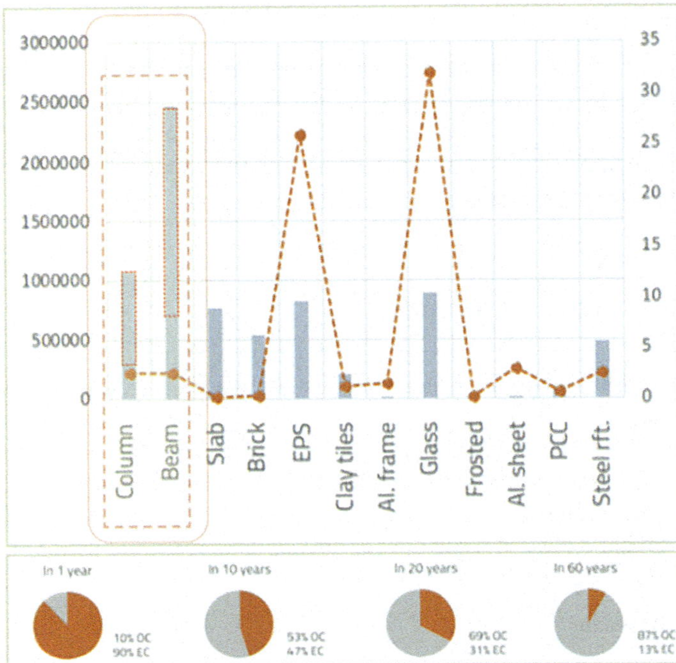

Fig. 7.8 Embodied carbon of elements using Recycled steel structural system and balance of carbon emissions over the life of the building

As the embodied carbon values are almost double the base iterations, it is clear that the use of recycled steel has a huge impact on carbon emissions. If the recycling rate of steel in India is documented, monitored, and expedited, it could have a significant impact on carbon emissions, particularly in the material phase.

Inferences

The embodied carbon emissions are 36% greater whereas the recycled steel content is lesser. Further effort in recycling steel can lead to larger carbon savings in the embodied carbon stages.

7.4.3 Summary

The changes are summarized below, and they indicate that bamboo-reinforced concrete is advantageous, but applicability needs to be investigated. Since the recycling rate in India for steel is lower than the global average, the embodied carbon values for recycled steel structures are high. If more metal recycling is encouraged and mandated, the carbon values may be lower (Table 7.6).

Table 7.6 Results of change in structural material

Parameter	Embodied carbon ($kgCO_2e/m^2$)	Operational carbon ($kgCO_2e/m^2/yr$)	EC change compared to conventional case (%)
Base case	421.6	73.15	−11
Conventional case	476	81	
Bamboo reinforcement	350	73.15	−27
Recycled steel	660	73.3	+38

The various iterations in line with the guidelines described in Chap. 4, have been simulated and the emissions quantified. These low-carbon alternatives provide a range of varying results which will be aggregated and discussed in upcoming chapters.

7.5 Conclusion

The design iterations applied to the baseline building as described in Fig. 7.1, indicate the following:

- **Using structural systems**—The use of agro-waste and industrial waste material in concrete mixes can result in the reduction of emissions by up to 12%. However, the mixes considered for the study were all under 40% replacement for GGBS and even lesser replacement percentages for the agro-mixes. If the cement mixes can be balanced with more additives without losing the stability and integrity of concrete structures, greater carbon reduction can be achieved. Despite the ongoing body of work in this sector, this is a potential area of study and implementation to reduce structural emissions, particularly in a developing country like India.
- **Replacement of steel and slabs**—The carbon emissions can be significantly reduced by 27% with these replacements. However, the stability and practicality of using bamboo and other natural materials must be further explored. And it needs consistent context-based research for inclusion in building design.
- If the structural design can be informed by the need to reduce embodied carbon emissions early in the design process, a significant reduction of building emissions can be achieved. Due to the nature of the early stage, exploring potential reductions through more structural measures is also possible in terms of building design and execution. Further in India, 2020 and 2070 are projected to reach 90.85 gigatons of carbon dioxide equivalent in the business-as-usual scenario, exceeding the carbon budget allocated for the entire country [7] and making the reduction of carbon emissions an urgent priority in the construction sector.

References

1. Madurwar MV, Ralegaonkar RV, Mandavgane S (2013) Application of agro-waste for sustainable construction materials: a review. Constr Build Mater 872–878
2. Kamble DR, Rajgure AV (2021) CO_2 emission assessment for rice husk ash in concrete on environmental approach. ResearchGate. https://www.researchgate.net/publication/350823893_CO2_EMISSION_ASSESSMENT_FOR_RICE_HUSK_ASH_IN_ CONCRETE_ON_ENVIRONMENTAL_APPROACH
3. Eziefula GU, Eziefula BI (2024) Incorporating bamboo leaf ash as a supplementary cementitious material in concrete: A review of the positive environmental and technical results. J Cleaner Produc. https://doi.org/10.1016/j.jclepro.2024.135832
4. Wang Z, Liu J, Liu X (2022) Study on the performance of alkali-activated materials using GGBS and MSWI bottom ash. Sustainability, 14(14):8783. https://doi.org/10.3390/su14148783
5. Ahmed S (2021) Bamboo reinforced concrete—strength, durability, selection & water absorption. Retrieved from Civil String—Get the knowledge. https://civilstring.com/bamboo-reinforced-concrete/
6. Puri S (2023) Recycling & CO_2 emission reduction in Indian Steel Sector. Retrieved from Linked-in: https://www.linkedin.com/pulse/recycling-co2-emission-reduction-indian-steel-sector-shivam-puri/
7. Economic Times. (2024, November 30). India's buildings sector to overshoot country's entire carbon budget: Report. Economic Times. https://energy.economictimes.indiatimes.com/news/renewable/indias-buildings-sector-to-overshoot-countrys-entire-carbon-budgetreport/108609597#~:text=%22In%20the%20business%2Das%2D,Buildings%20Sector%20Towards%20a%20Net%2D

Chapter 8
Envelope Material Iterations

8.1 Introduction

Building envelopes are among the systems that can have the largest impact on a building's carbon emissions. This is due to two major factors:

1. The volume or quantity of envelope materials is significantly more than most of the other materials. This is also seen in Chap. 5.
2. Any change in the envelope material is likely to affect the embodied and operational carbon as the envelope of the building forms the thermal barrier between the structure and the environment.

The challenge with Building Envelopes is that they are governed by multiple parameters. It includes shading devices, glazing properties, Window wall ratio (WWR), and many others. As these are predominantly passive design measures, these interventions are heavily dependent climate, context, and location of the proposed building. This section follows the concept of Chap. 6 and describes the impact of selected envelope iterations on carbon emissions. These iterations are shown in Table 8.1.

As discussed, the alternatives illustrated in Fig. 8.1 are not comprehensive. There may be various materials and types of systems available in a given local context. The most accessible source to reduce transport emissions should be considered. This theoretical model, however, assumes a series of 7 iterations for the purpose of envelope material impact study and derivation of comparative inferences. The following sections describe the boundaries, assumptions, and results of the guidelines.

S. Mohan and Vijayalaxmi J., *Embodied and Operational Carbon in Buildings*,
SpringerBriefs in Architectural Design and Technology,
https://doi.org/10.1007/978-981-97-7187-5_8

Table 8.1 Methodology to apply the design guidelines with respect to context

<table>
<tr><th>S.no</th><th>Parameters</th><th>Guideline</th><th>Building type</th><th>Stage</th><th>Considerations</th><th>Application</th></tr>
<tr><td rowspan="4">1</td><td rowspan="4">Alternative materials used</td><td rowspan="4">Use of responsibly sourced natural materials like strawbales, cellulose insulation, rammed earth, CSEB, etc., and waste-based materials like Fly ash, PFA, GGBS, etc</td><td rowspan="4">New buildings/ retrofit existing buildings</td><td rowspan="4">Concept design and design development</td><td rowspan="2">Key considerations include specification and mixes of materials and volume of material used/ replaced in the building</td><td>Use of bio-based materials (Porotherm blocks, etc.)</td></tr>
<tr><td>Use of natural materials (Rammed earth, CSEB, etc.)</td></tr>
<tr><td rowspan="2">Locally produced agricultural and industrial waste can be used as cement replacement in green concrete for envelope materials</td><td>Use of industrial waste, and salvaged materials (FAlg, AAC blocks, etc.)</td></tr>
<tr><td>Use of roof alternatives (Green roof, filler slab, etc.)</td></tr>
</table>

8.2 Alternative Material Iterations

Alternative materials for low embodied carbon are considered under three categories.

- Natural materials.
- Biomaterials.
- Salvaged materials.

All the material iterations are involved.

1. Natural materials
    - Rammed Earth.
    - CSEB.
    - Unfired straw earth blocks.
2. Bio-based materials
    - Porotherm bricks.
    - AAC block.
3. Salvaged materials

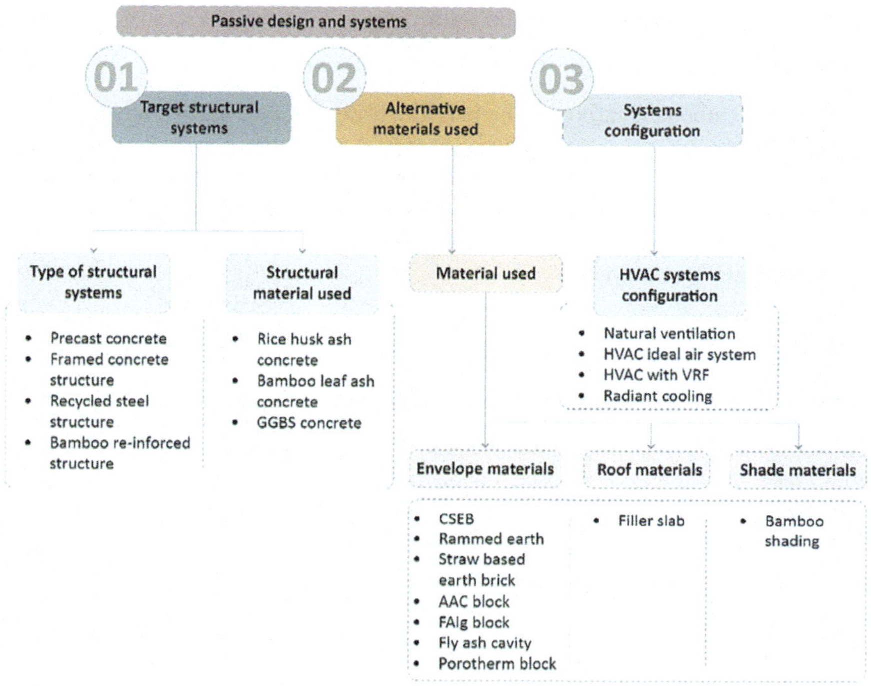

Fig. 8.1 Design-based iterations considered

- Fly ash brick.
- Fly ash cavity block.

8.2.1 Rammed Earth

Natural materials are the most easily available renewable material, and hence they have very low embodied carbon.

The walls or envelope material forms the largest volume of building material. A low-carbon material used for walls could significantly reduce carbon emissions.

Key considerations:

- Rammed earth additionally has excellent thermal mass, which impacts the operational energy usage as well.
- The thickness is assumed to be 0.2 m. It can be used in multi-story constructions (load-bearing or non-load-bearing) and can be engineered to achieve high strength and be reinforced in a similar manner to concrete.

Embodied and operational carbon data:

For embodied carbon of the building with rammed earth:

- The total embodied carbon of the building using rammed earth walls is 4444209 $kgCO_2e$.
- These carbon emissions are normalized per building area **398.6 $kgCO_2e/m^2$**.
- These normalized embodied carbon emissions are then divided by the lifespan of the building (60 years) to calculate **annual embodied carbon**, which in this case, is **6.64 $kgCO_2e/m^2/year$**.

The embodied carbon values are also represented in Fig. 8.2, with the red line indicating the difference in embodied carbon from the base case of the building.

For operational carbon of the building with rammed earth:

The same methodology, discussed in Chap. 7 using Rhino and Grasshopper software with Open Studio engine is used for energy analysis and to arrive at operational carbon emissions of the proposed iteration. Apart from the material properties of the envelope, the other parameters for the study remain the same as the baseline.

- The EUI of the proposed building is 98.03 $KWh/m^2/year$.
- The **operational carbon** using India's CE factor is **79.4 $kgCO_2e/m^2/year$**.

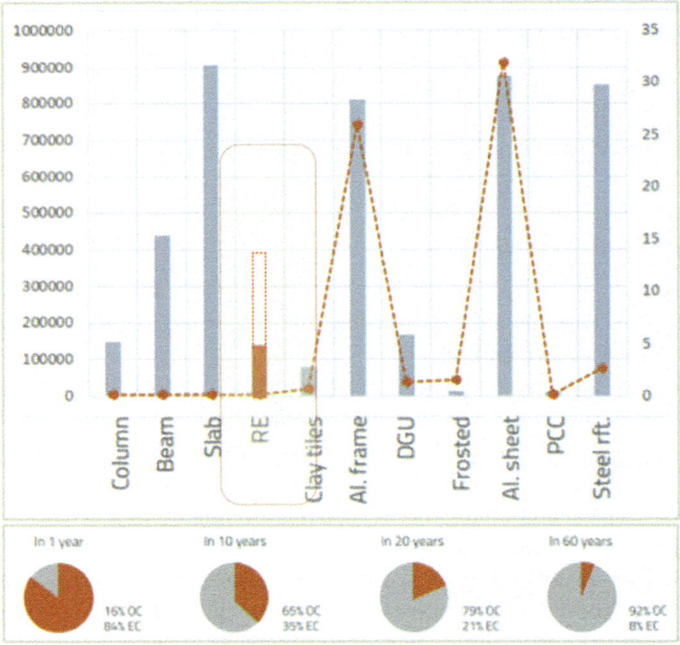

Fig. 8.2 Embodied carbon of elements using rammed earth walls and balance of carbon emissions over the life of the building

The embodied carbon values of rammed earth walls are lower than the base case, which has a double brick cavity wall. As a result, the operational carbon emissions are also higher as compared to the base case. This indicates that the change in wall materials has an impact on both embodied and operational carbon emissions.

8.2.2 CSEB

Compressed stabilized earth blocks (CSEB) are manufactured from local soil mixed/ stabilized with a small amount of cement (up to 5%), sand, and water. Greenhouse gas emissions are significantly reduced as these blocks are not fired in kilns as done in the case of burnt clay bricks.

Key considerations:

- The amount of cement used for stabilization varies with context, nature of soil characteristics, compressibility, cohesion, and other specific factors.
- The cement stabilization used for the soil blocks in this study is 5%.

Embodied and operational carbon data:

For embodied carbon of the building with CSEB:

- The total embodied carbon of the building with CSEB envelope is 4557266 $kgCO_2e$.
- These carbon emissions are normalized per building area **408.7 $kgCO_2e/m^2$**.
- The normalized embodied carbon using CSEB is divided by the lifespan of the building (60 years) to calculate annual embodied carbon, which is **6.81 $kgCO_2e/m^2$/year**.

The embodied carbon values are also represented in Fig. 8.3, with the red line indicating the difference in embodied carbon from the base case of the building.

For operational carbon of the building with CSEB:

The same methodology, discussed in Chap. 7 using Rhino and Grasshopper software with Open Studio engine is used for energy analysis and to arrive at operational carbon emissions of the proposed iteration. Apart from the material properties of the envelope, the other parameters for the study remain the same as the baseline.

- The EUI of the proposed building is 98.03 KWh/m^2/year.
- The **operational carbon** using India's CE factor is 79.4 $kgCO_2e/m^2$/year.

Inferences:

The embodied carbon emissions of CSEB are 3% lower than the baseline carbon emissions and the operational carbon emissions are 8% greater than baseline carbon emissions with a double brick cavity wall. This indicates that insulated walls are better

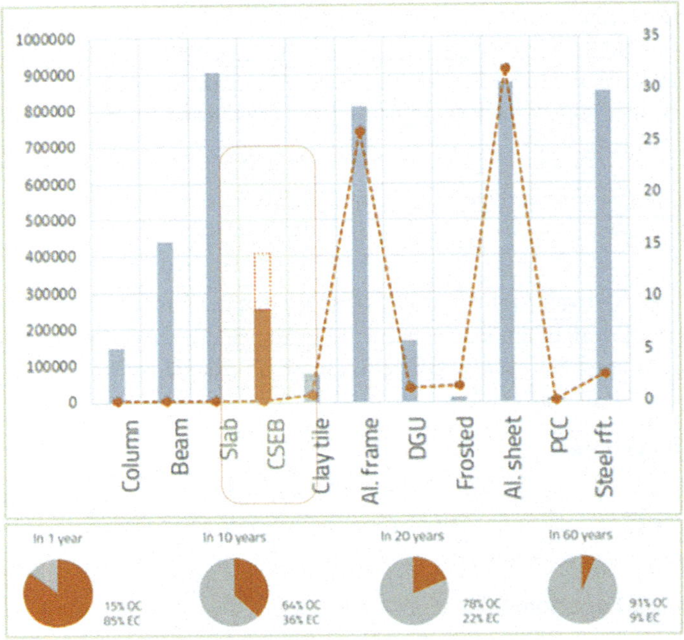

Fig. 8.3 Embodied carbon of elements using CSEB walls and balance of carbon emissions over the life of the building

performing in terms of operational emissions compared to single-layered natural materials. While the embodied carbon of CSEB is better compared to a baseline.

8.2.3 Straw-Based Unfired Earth Block

Straw fiber mixed earth-based unfired blocks are a combination of natural and bio-based materials. These can be stabilized with cement or lime in the mix. The blocks can also be solid unfired earthen blocks or hollow earthen blocks.

Key considerations:

- Key considerations include the compressive strength of the reinforced earth blocks. The straw fibers used in the earth blocks are at a ratio of less than 2.5% in volume.
- It is not a load-bearing material and requires good testing and applicability on a proposed building.

Embodied and operational carbon data:

For embodied carbon of the building with Straw-based brick:

- The total embodied carbon of the building using straw-based brick for the envelope material is 4351309 kgCO_2e.
- These carbon emissions are normalized per building area **390.25 kgCO_2e/m^2**.
- The normalized embodied carbon is divided by the lifespan of the building (60 years) to calculate **annual embodied carbon**, which is **6.5 kgCO_2e/m^2/year**.

The embodied carbon values are also represented in Fig. 8.4, with the red line indicating the difference in embodied carbon from the base case of the building.

For operational carbon of the building with Straw-based brick:

The same methodology, discussed in Chap. 7 using Rhino and Grasshopper software with Open Studio engine is used for energy analysis and to arrive at operational carbon emissions of the proposed iteration.

- The EUI of the proposed building is 90.104 KWh/m^2/year.
- The **operational carbon** using India's CE factor is **72.98 kgCO_2e/m^2/year**.

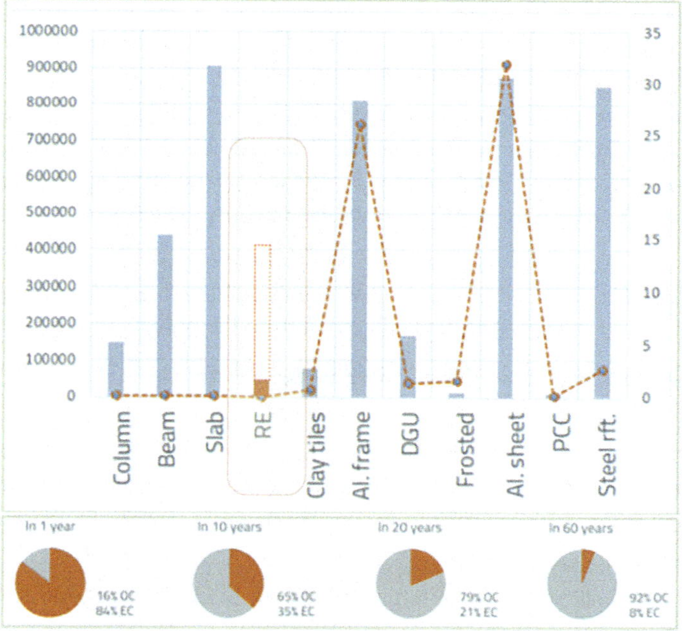

Fig. 8.4 Embodied carbon of elements using Straw-based earth block walls and balance of carbon emissions over the life of the building

Inferences:

The use of straw fiber-based unfired earth blocks reduces the embodied carbon emissions from the base case by 8%, though the percentage of straw fibers is less. The operational carbon emissions are also reduced by 1%, indicating a better low-carbon iteration case for envelope materials.

8.2.4 AAC Block

AAC blocks have revolutionized the construction industry by offering a sustainable, cost-effective, and safe alternative to traditional building materials. They are low density, have self-insulation, are fire resistant, flexible, and are low-carbon materials.

Key considerations:

- The AAC blocks are considered with 0.2 m thickness. They are non-load bearing and used as a filler material for the building envelope.
- They are made from natural materials and are recyclable. They can be crushed and reused and are considered sustainable.

Embodied and operational carbon data:

For embodied carbon of the building with AAC Block:

- The total embodied carbon of the building using AAC block is 4566906 $kgCO_2e$.
- These emissions are normalized per building area **409.06** $\mathbf{kgCO_2e/m^2}$.
- The normalized embodied carbon is divided by the lifespan of the building (60 years) to calculate annual **embodied carbon**, which is **6.83** $\mathbf{kgCO_2e/m^2/year}$.

The embodied carbon values are also represented in Fig. 8.5 with the red line indicating the difference in embodied carbon from the base case of the building.

For operational carbon of the building with AAC Block:

The same methodology, discussed in Chap. 7 using Rhino and Grasshopper software with Open Studio engine is used for energy analysis and to arrive at operational carbon emissions of the proposed iteration. Apart from the material properties of the envelope, the other parameters for the study remain the same as the baseline.

- The EUI of the proposed building is 91.227 $KWh/m^2/year$.
- The operational carbon using India's CE factor is **73.9** $\mathbf{kgCO_2e/m^2/year}$.

Inferences:

The use of AAC blocks as wall materials reduces the embodied carbon emissions from the base case by 3%. The operational carbon emissions are increased by 1% indicating that cavity walls perform better for operational emissions, though embodied emissions are higher.

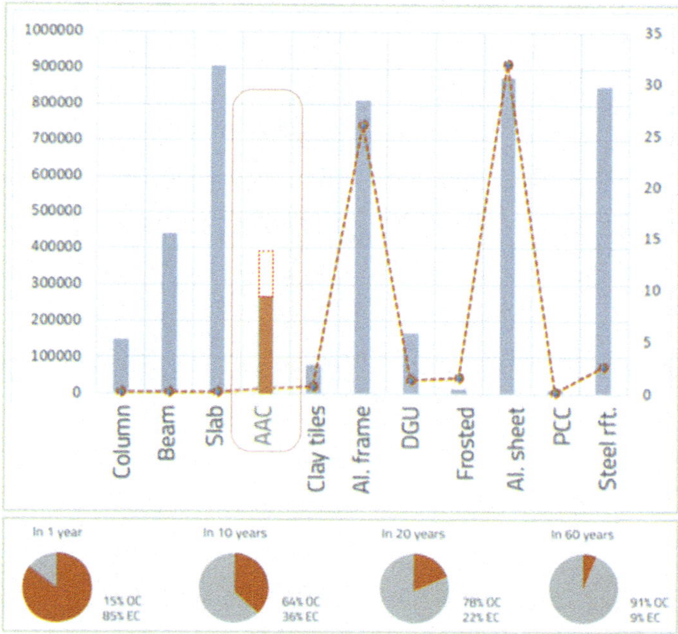

Fig. 8.5 Embodied carbon of elements using AAC block walls and balanced carbon emissions over the life of the building

8.2.5 *Fly Ash Brick with a Cavity in Between*

Indian coal is low grade and has a high ash content (30–45%) compared to imported coals, so large quantities of fly ash are generated. In India, the utilization of fly ash by the cement industry amounts to around 60.11 million-ton of the total production in the 2018–19 calendar year.

Key considerations:

- The cavity wall is made of fly ash bricks of 0.2 m thickness. The cavity is filled with insulation materials such as expanded polystyrene (EPS) and used in the envelope of the proposed building.
- The composition, density, and finish of the fly ash blocks, will have an impact on the carbon emissions.

Embodied and operational carbon data:

For embodied carbon of the building with Fly ash brick:

- The total embodied carbon of the building using fly ash brick with a cavity is 473850 $kgCO_2e$.
- These carbon emissions are normalized per building area of **425 $kgCO_2e/m^2$**.

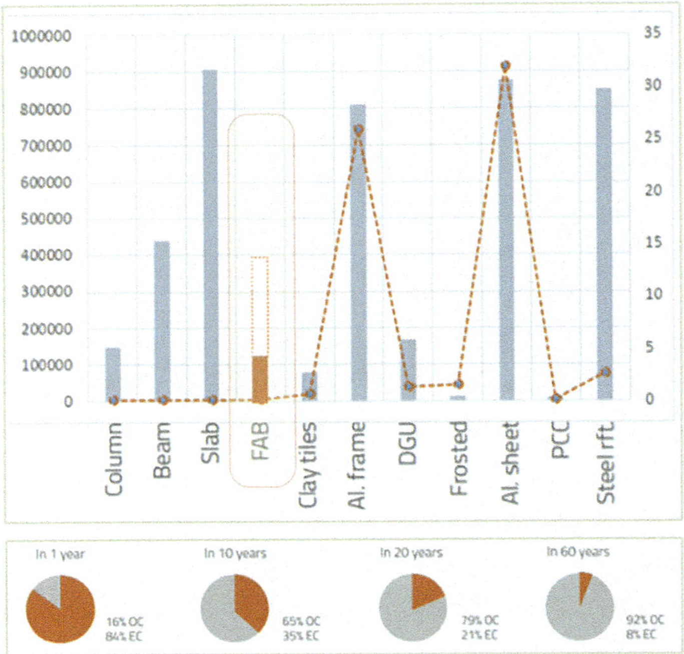

Fig. 8.6 Embodied carbon of elements using Fly ash cavity walls and balance of carbon emissions over the life of the building

- The normalized embodied carbon is divided by the lifespan of the building (60 years) to calculate **annual embodied carbon**, which is **7.08 kgCO_2e/m^2/year**.

The embodied carbon values are also represented in Fig. 8.6 with the red line indicating the difference in embodied carbon from the base case of the building.

For operational carbon of the building with Fly ash brick:

The same methodology, discussed in Chap. 7 using Rhino and Grasshopper software with Open Studio engine is used for energy analysis and to arrive at operational carbon emissions of the proposed iteration. Apart from the material properties of the envelope, the other parameters for the study remain the same as the baseline.

- The EUI of the proposed building is 90.1 KWh/m^2/year.
- The operational carbon using India's CE factor is **72.98 kgCO_2e/m^2/year**.

Inferences:

The use of fly ash cavity walls has reduced embodied emissions by 2% compared to the base case of burnt brick cavity walls. The volume of material used in the envelope is similar to the base case. The operational carbon emissions are also 2% below the baseline, due to the thermal performance which is better due to the presence of the cavity wall.

8.2.6 Porotherm Block

Porotherm bricks are clay structural walling devices with high strength and thermal potency. The raw materials used are industrial waste or recycled materials. Thus, they save on natural materials and are therefore low carbon. Some manufacturers of Porotherm bricks are also IGBC rated and have been used for LEED and GRIHA projects as well.

Key considerations:

- The thickness of the porotherm block used is 0.2 m. They are made of natural clay, rice husk, coal ash, and granite slurry.
- It allows insulation of buildings due to the nature of the material but requires skilled labor and has limited availability.

Embodied and operational carbon data:

For embodied carbon of the building with Porotherm block:

- The total embodied carbon of the building using the Porotherm block is 4422299 $kgCO_2e$.
- These carbon emissions are normalized per building area **396.61 $kgCO_2e/m^2$**.
- The normalized embodied carbon is divided by the lifespan of the building (60 years) to calculate **annual embodied carbon**, which is **6.61 $kgCO_2e/m^2$/ year**.

The embodied carbon values are also represented in Fig. 8.7 with the red line indicating the difference in embodied carbon from the base case of the building.

For operational carbon of the building with Porotherm block:

The same methodology, discussed in Chap. 7 using Rhino and Grasshopper software with Open Studio engine is used for energy analysis and to arrive at operational carbon emissions of the proposed iteration. Apart from the material properties of the envelope, the other parameters for the study remain the same as the baseline.

- The EUI of the proposed building is 100.3 KWh/m^2/year.
- The operational carbon using India's CE factor is **81.24 $kgCO_2e/m^2$/year**.

Inferences:

The use of Porotherm block walls has reduced embodied emissions by 6% compared to the base case of burnt brick cavity walls. However, the operational carbon emissions are greater than the baseline by 10%, indicating that cavity walls have lower carbon emissions compared to Porotherm blocks.

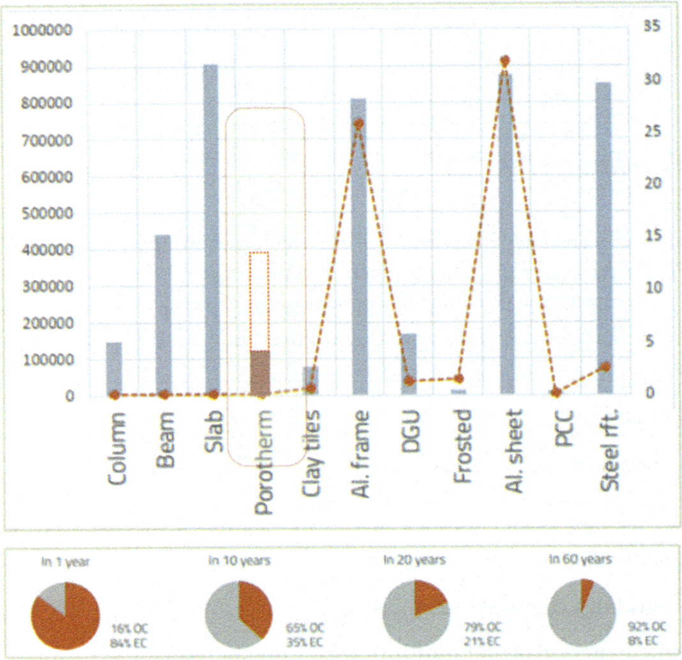

Fig. 8.7 Embodied carbon of elements using Porotherm block walls and balance of carbon emissions over the life of the building

8.2.7 *Summary*

The operational carbon value of brick walls (conventional) is **85.98 kgCO$_2$e/ sq.m/ year**.

The operational carbon value of the brick wall (existing) is **73.15 kgCO$_2$e/ sq.m/ year** (Fig. 8.8).

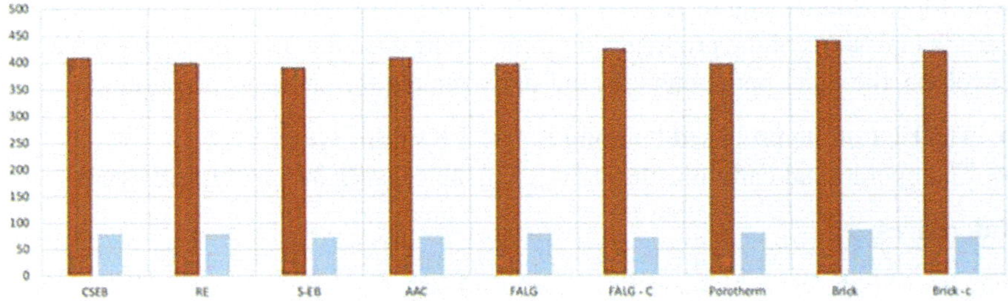

Fig. 8.8 Comparison of walling material OC and EC

Table 8.2 Summary of alternate envelope materials used

Parameter	Embodied carbon ($kgCO_2e$/sq.m)	Operational carbon ($kgCO_2e$/ sq.m/yr)	EC change compared to conventional case (%)	OC change compared to conventional case (%)
Base case	421.6	73.15	−11	−10
Conventional case	476	81		
CSEB	408.7	79.5	−15	−2
Rammed earth	398.6	79.4	−17	−2
Straw-based brick	390.2	72.98	−19	−10
AAC block	409.6	73.9	−14	−9
FaLG	397.2	79.4	−17	−2
FaLG with cavity	425	72.9	−11	−10
Porotherm	396	81.2	−17	0

The operational carbon value of the FALG cavity is 72.98 kg CO_2 e/ sq.m/year. This achieves a 9% reduction from conventional methods.

The embodied carbon value of brick wall (conventional) is 441 kg CO_2 e/ sq.m

The embodied carbon value of cavity brick wall (existing) is 421 kg CO_2 e/ sq.m

The embodied carbon value of straw-based earth is 390.25 kg CO_2e/sq.m

This achieves a reduction of 12% from conventional using alternative walling materials.

Cavity walls technique and insulation can be used to arrive at a further reduction in the values (Table 8.2).

The table indicates that cavity-based walls, while having bigger embodied carbon have lower operational emissions. Single-leaf walls or walls with no cavity have lower embodied emissions and higher operational emissions.

8.3 Alternatives to Application on Walls

8.3.1 Filler Slab

The slab is the element that contributes the largest to carbon emissions. Substituting the concrete slab with the re-iforce ment used in the base case is one of the iterations used.

Key considerations:

- The fundamental concept behind the filler slab is that an RCC slab does not need concrete located in the slab's lower half for its construction.
- A lightweight, inexpensive filler material like clay pots, Mangalore tiles, etc., is used to replace this concrete component.
- The thickness of the slab is around 0.15 m to 0.2 m. However, they require careful design and engineering and skilled labor for applicability.
- The loading on the slab must be considered for any iterations on roof material.

Embodied and operational carbon data:

For embodied carbon of the building with filler slabs:

- The total embodied carbon of the building with filler slabs is 4087164 $kgCO_2e$.
- These emissions are normalized per building area **366.56** $\mathbf{kgCO_2e/m^2}$.
- The normalized embodied carbon is divided by the lifespan of the building (60 years) to calculate **annual embodied carbon**, which is **6.1** $\mathbf{kgCO_2e/m^2/year}$.

The embodied carbon values are also represented in Fig. 8.9 with the red line indicating the difference in embodied carbon from the base case of the building.

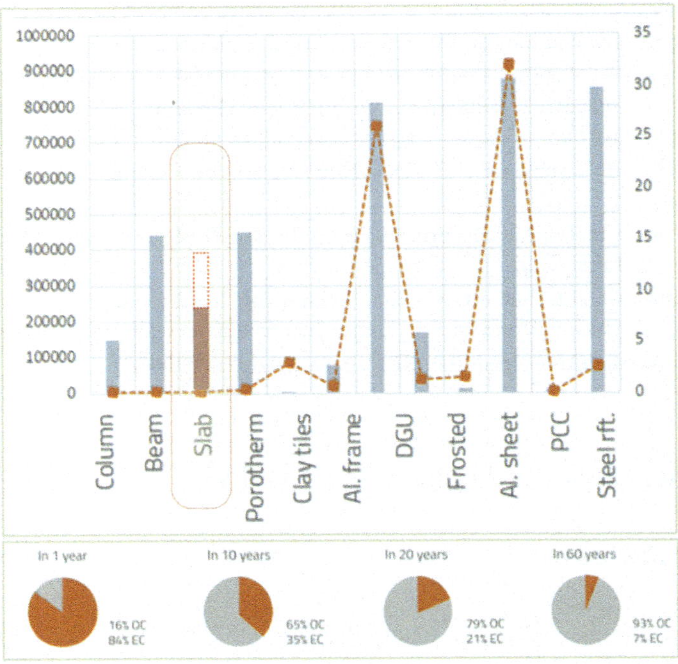

Fig. 8.9 Embodied carbon of elements using a Filler slab and balance of carbon emissions over the life of the building

For operational carbon of the building with filler slabs:

The same methodology, discussed in Chap. 7 using Rhino and Grasshopper software with Open Studio engine is used for energy analysis and to arrive at operational carbon emissions of the proposed iteration. Apart from the material properties of the envelope, the other parameters for the study remain the same as the baseline.

- The EUI of the proposed building is 89.1 $KWh/m^2/year$.
- The **operational carbon** using India's CE factor is **72.17 $kgCO_2e/m^2/year$**.

Inferences:

The operational carbon has not been affected by the filler slab; however, embodied carbon emissions show a 12% reduction due to the reduction in the amount of concrete required for the slab.

8.3.2 Bamboo Shading

According to the base case analyzed in Chap. 6, the aluminum external shades and window frames are large contributors to carbon emissions. This iteration aims to replace the shading material and systems with natural or waste-based materials.

Key considerations:

- Exposed bamboo elements need to be treated to be used externally for water-proofing and exposure to other elements.
- The composition of the building and associated joints and construction must be treated carefully.

Embodied and operational carbon data:

For embodied carbon of the building with bamboo shading:

- The total embodied carbon of the building with bamboo shading is 3881363.8 $kgCO_2e$.
- These carbon emissions are normalized per building area **349 $kgCO_2e/m^2$**.
- The normalized embodied carbon is divided by the lifespan of the building (60 years) to calculate **annual embodied carbon**, which is **5.8 $kgCO_2e/m^2/year$**.

The embodied carbon values are also represented in Fig. 8.10 with the red line indicating the difference in embodied carbon from the base case of the building.

For operational carbon of the building with bamboo shading:

The same methodology, discussed in Chap. 7 using Rhino and Grasshopper software with Open Studio engine is used for energy analysis and to arrive at operational carbon emissions of the proposed iteration. Apart from the material properties of the envelope, the other parameters for the study remain the same as the baseline.

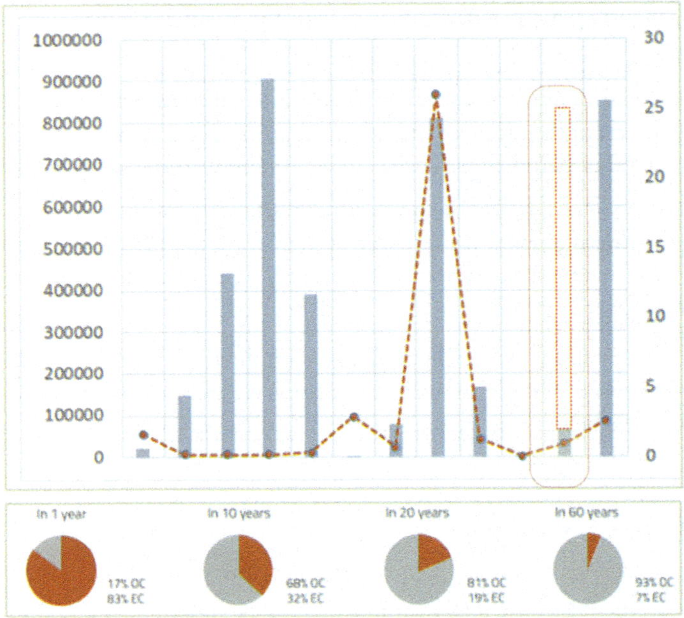

Fig. 8.10 Embodied carbon of elements using Bamboo shading devices and balanced carbon emissions over the life of the building

- The EUI of the proposed building is 90.3 KWh/m^2/year.
- The operational carbon using India's CE factor is **73.15 kgCO$_2$e/m^2/year**.

Inferences:

The operational carbon has not been affected by the bamboo slab; however, embodied carbon emissions show a 14% reduction due to the replacement of aluminum in the base case.

8.3.3 Summary

It is seen that using alternative materials for elements like roofs, shading devices, glass, etc. can also benefit immensely in terms of carbon emissions, particularly embodied emissions.

They do not have a big impact on operational emissions or comfort as seen in the simulation. So, alternative areas for the application of low-carbon materials can also be looked at and addressed clearly by the rating systems (Table 8.3).

Table 8.3 Results of change in alternative material

Parameter	Embodied Carbon ($kgCO_2e$/sq.m)	Operational Carbon ($kgCO_2e$/sq.m/yr)	EC change compared to conventional case (%)	OC change compared to conventional case (%)
Base case	421.6	73.15	−11	−10
Conventional case	476	81		
Filler slab	366.56	72.1	−26	−12
Bamboo shading	349	73.15	−27	−10

8.4 Conclusions

The design iterations applied to the baseline building, indicate the following:

- **Alternatives to wall materials**—A change of wall materials can show up to a 20% reduction in embodied carbon and a 10% reduction in operational emissions. However, this parameter is highly context specific and can be used very effectively to equally impact embodied and operational carbon emissions of a building.
- **Alternative other materials**—Effective application of alternative materials like bamboo to carbon hotspots can lead to reductions of up to 27%. Apart from the context specificity, this parameter is also design-specific and can be addressed efficiently at an early stage to avoid carbon-intensive materials.
- The study demonstrates that thoughtful material selection can reduce carbon emissions and make a significant impact on achieving net zero targets.
- The changes in envelope material are also dependent on other variables like the typology of the building, service life, and retrofit or reuse of the buildings. As it is a flexible element, more exploration and investigation into low-carbon material possibilities can be implemented, particularly at an early stage of building development.

Chapter 9
HVAC System Iterations

9.1 Introduction

Along with embodied carbon, operational energy and carbon emissions, form more than 80% of the building's carbon emissions. Therefore, a large focus is laid on energy efficiency, decarbonization of the grid, use of renewable energy, increase of service life, and other measures to reduce the operational emissions of a building.

Of these, Heating, Ventilation, and Air-Conditioning (HVAC) systems are key contributors to carbon emissions during the operating life. As India is a tropical country, the increasing use of air conditioners could lead to a rise of at least 120 million metric tons in emissions by 2050 (Times of India, 2023). Therefore, it is essential to explore the reduction of carbon emissions by exploring alternatives to current HVAC systems.

A few of the options are described in this chapter. The quantification of these iterations highlights the impact and is used to discuss feasible strategies for carbon-conscious design and decarbonization of buildings (Table 9.1).

For systems and equipment:

- HVAC has the maximum energy share among systems in office buildings and hence has been considered to check reduction in operational carbon and associated embodied carbon balance in buildings.

The alternatives illustrated in Fig. 9.1 are not comprehensive. There may be various materials and types of systems available in a given local context. The most accessible source to reduce transport emissions should be considered. This theoretical model, however, assumes a series of 4 HVAC iterations for the purpose of impact study and derivation of comparative inferences.

S. Mohan and Vijayalaxmi J., *Embodied and Operational Carbon in Buildings*,
SpringerBriefs in Architectural Design and Technology,
https://doi.org/10.1007/978-981-97-7187-5_9

Table 9.1 Methodology to apply the design guidelines with respect to context

S.no	Parameters	Guideline	Building type	Stage	Considerations	Application
1	Type of system	Use of different types of configurations like VAV system, VRF radiant cooling, etc.	New buildings/ retrofit existing buildings	Design development	Type of air conditioning systems and efficiency of the system used	Use of VAV, VRF, Radiant cooled systems, etc.

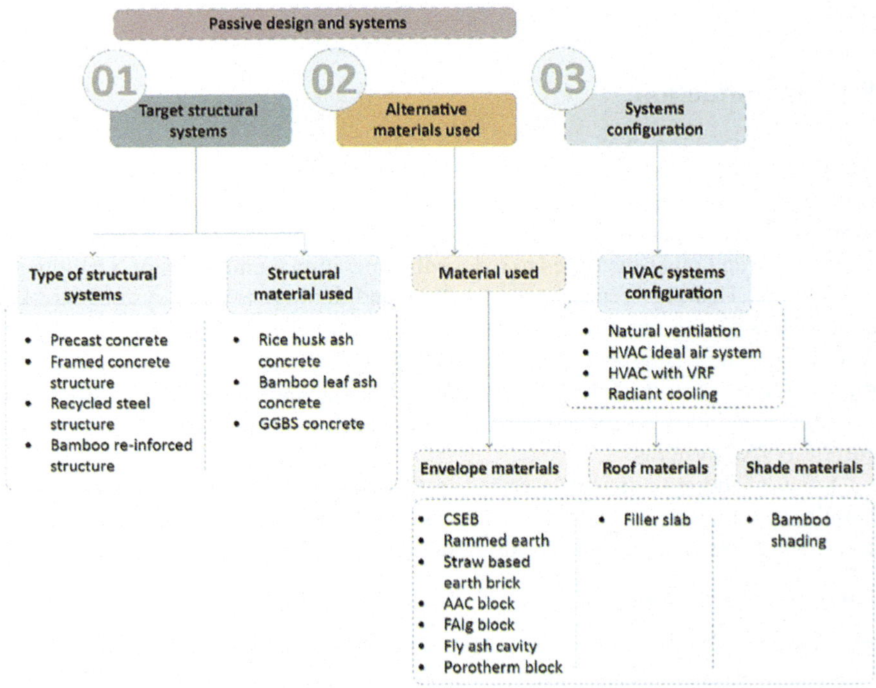

Fig. 9.1 Design-based iterations considered

9.2 System Iterations

As discussed, earlier, HVAC systems have a big effect on both the operational carbon emissions as well as the comfort of the occupant.

This is evaluated using four common HVAC system configurations.

1. VAV (Variable Air Volume) system which is as existing (base case).
2. VRF (Variable Refrigerant Flow) system.

3. Air-based cooling systems.
4. Radiant cooling, and
5. Natural ventilation.

The parameters for evaluating comfort are average operative temperature for comparative purposes and percentage of discomfort hours.

9.2.1 VRF System

Modern VRF systems provide some major advantages, such as zoning, individual temperature control, minimized ductwork, excluding the need for secondary fluids (chilled-water or hot-water distribution), and associated costs. These can be used to replace the base case VAV system.

Key considerations:

- The type of refrigerant used plays a key role in carbon emissions from HVAC systems.
- As the system could be electrified, local electricity sources could affect emissions. If renewable sources are used, the carbon emissions could be reduced.
- The conditioning of areas, setpoints, and occupancy schedules will also form key parameters that determine carbon emissions. In this case, an office occupancy schedule is used.

Embodied and operational carbon data:

For embodied carbon of the building with VRF system:

- The total embodied carbon of the building using the VRF system is 4700404 $kgCO_2e$.
- These emissions are normalized per building area **421.6 $kgCO_2e/m^2$**.
- The normalized embodied carbon is divided by the lifespan of the building (60 years) to calculate **annual embodied carbon**, which is **7.03 $kgCO_2e/m^2/year$**.

The embodied carbon values are very similar to the base case as seen in Fig. 9.2.

For operational carbon of the building with VRF system:

The same methodology, discussed in Chap. 7 using Rhino and Grasshopper software with Open Studio engine is used for energy analysis and to arrive at operational carbon emissions of the proposed iteration. The HVAC system definition is changed for this option, while other parameters are maintained the same as the base case.

- The EUI of the proposed building is 62.727 $KWh/m^2/year$.
- The operational carbon using India's CE factor is **50.81 $kgCO_2e/m^2/year$**.

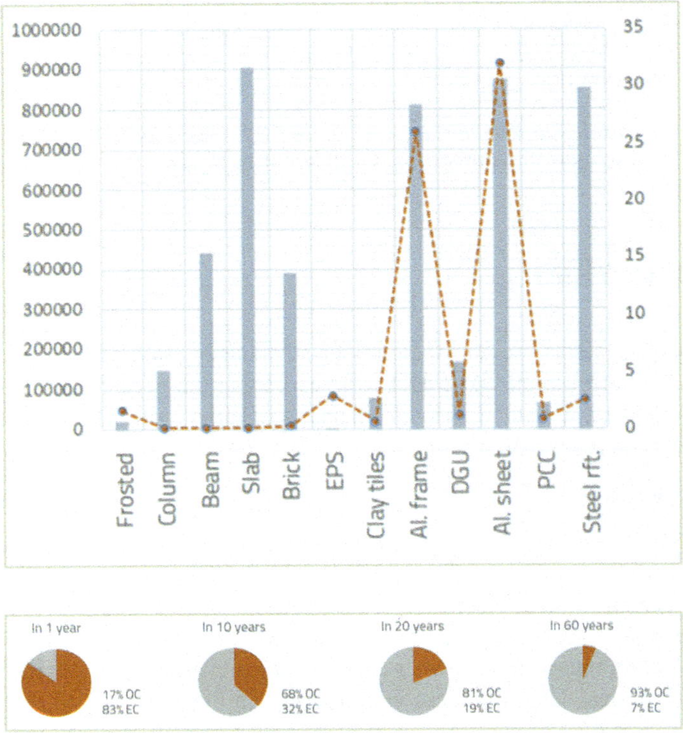

Fig. 9.2 Embodied carbon of elements using HVAC VRF system and balance of carbon emissions over the life of the building

Inferences:

System materials are a less significant contributor to embodied carbon emissions and are hence not considered. The operational carbon of the VRF system is low carbon, but GHG emissions are dependent on the type of refrigerant used as well. VRF systems are more environmentally friendly energy solutions that can greatly reduce carbon emissions from buildings.

9.2.2 Radiant Cooling

A radiant cooling system cools a floor or a ceiling by absorbing the heat radiated to the space. This can be done by using chilled slabs or ceiling panels. The coils are embedded in the structure, and they carry chilled water to cool the mass. These systems are less energy intensive than traditional HVAC systems and can be an effective low-carbon alternative in certain climates.

Key considerations:

- The mass of the building is a key consideration. Concrete structures with radiant cooling can increase thermal mass, thereby increasing lag in heat transfer.
- The temperature of the chilled water is a key consideration. It can be chilled just enough to provide comfort and is flexible to the context and climate.

Embodied and operational carbon data:

For embodied carbon of the building with radiant cooling:

- The total embodied carbon of the building with a radiant cooling system is 4700404 kgCO_2e.
- These emissions are normalized per building area **421.6 kgCO_2e/m^2**.
- The normalized embodied carbon is divided by the lifespan of the building (60 years) to calculate **annual embodied carbon**, which is **7.03 kgCO_2e/m^2/year**.

The embodied carbon values are very similar to the base case as seen in Fig. 9.3.

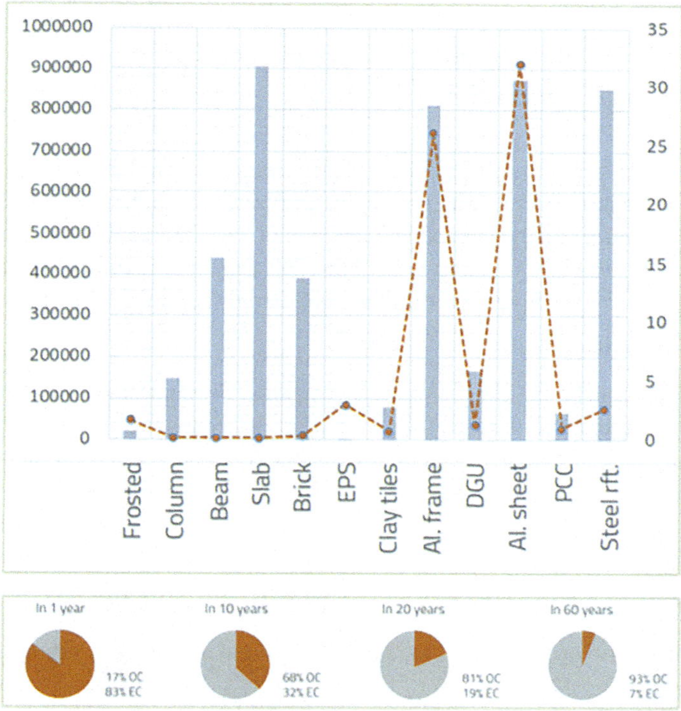

Fig. 9.3 Embodied carbon of elements using a radiantly cooled system and balance of carbon emissions over the life of the building

For operational carbon of the building with radiant cooling:

The same methodology, discussed in Chap. 7 using Rhino and Grasshopper software with Open Studio engine is used for energy analysis and to arrive at operational carbon emissions of the proposed iteration. The HVAC system definition is changed for this option, while other parameters are maintained the same as the base case.

- The EUI of the proposed building is 55 KWh/m^2/year.
- The operational carbon using India's CE factor is **44.55 kgCO_2e/m^2/year**.

Inferences:

The installation of a radiant cooling system may also lead to a significant reduction in forced-air system components and ductwork, though the corresponding impact on overall embodied carbon is minimal, as this project considers only civil work for embodied carbon. Compared to traditional systems, radiant cooling is flexible, allowing a greater setpoint to maintain the same level of comfort. This can lead to a significant reduction in energy consumption and carbon footprint as indicated in the charts.

9.2.3 Natural Ventilation

Natural ventilation is traditionally the least intrusive ventilation mode possible. However, India is a tropical country, where temperatures soar high at certain times of the year. Depending on the context and location of the building, it may or may not be feasible to resort purely to natural ventilation. (Vijayalaxmi, 2009). A hybrid mode with time periods of natural and mechanical ventilation might be more appropriate.

Key considerations:

- The design of the building, in terms of passive design outlined in Chap. 5 is a major consideration, for the thermal comfort of occupants.
- Further, local climate and geographical context play a key role in the movement of wind for natural ventilation.

Embodied and operational carbon data:

For embodied carbon of the building with Natural ventilation:

- The total embodied carbon of the building without any HVAC system is 4700404 kgCO_2e.
- The embodied carbon emissions normalized per building area **421.6 kgCO_2e/m^2**.
- The normalized embodied carbon is divided by the lifespan of the building (60 years) to calculate **annual embodied carbon**, which is **7.03 kgCO_2e/m^2/year**.

For operational carbon of the building with Natural ventilation:

The same methodology, discussed in Chap. 7 using Rhino and Grasshopper software with Open Studio engine is used for energy analysis and to arrive at operational carbon emissions of the proposed iteration.

- The EUI of the proposed building is 46.1 KWh/m^2/year.
- The operational carbon using India's CE factor is **37.31 kgCO$_2$e/m^2/year** (Fig. 9.4).

Inferences:

Though natural ventilation appears to have the least operating emissions, the thermal comfort of the occupants is a key factor in decision-making. Depending on the context and climate, this may vary for different buildings. However, if natural ventilation can be employed whenever possible, it will reduce the operating and embodied emissions of the proposed building.

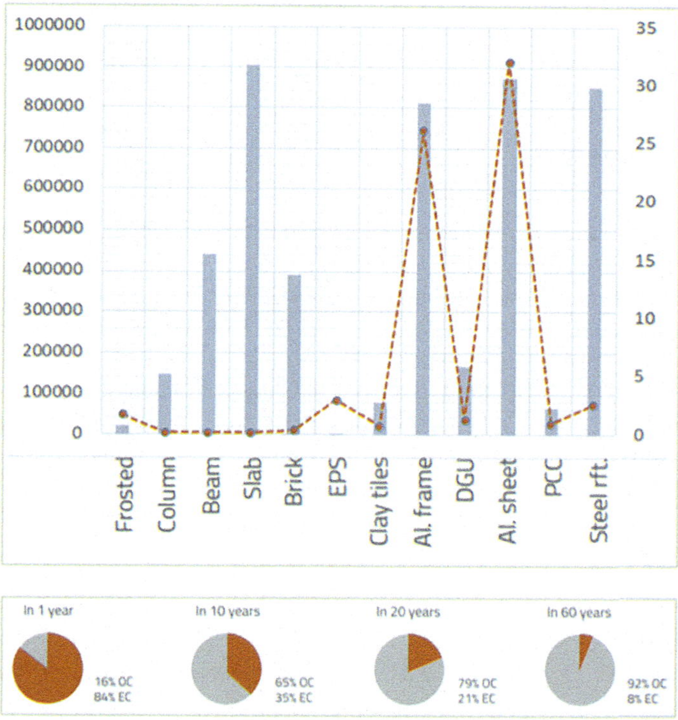

Fig. 9.4 Embodied carbon of elements using natural ventilation and balance of carbon emissions over the life of the building

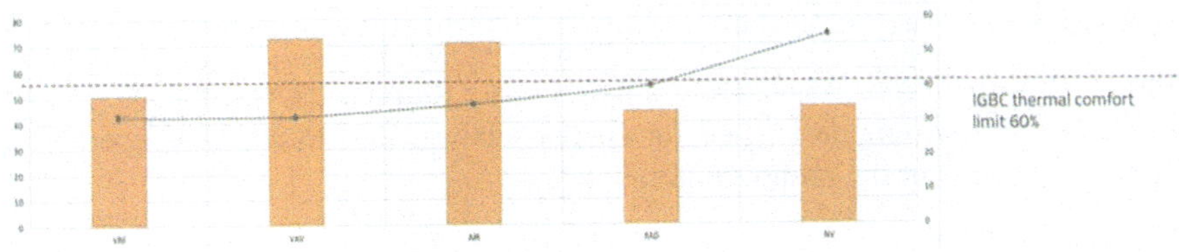

Fig. 9.5 Comparison of systems

Table 9.2 Results of change in HVAC systems

Parameter	Embodied carbon ($kgCO_2e$/sq.m)	Operational carbon ($kgCO_2e$/sq.m/yr)	OC Change compared to conventional case (%)
Base case	421.6	73.15	−10
Conventional case	476	81	
HVAC with VRF	421.6	50.81	−38
Radiant cooling	421.6	44.55	−45
Air-based HVAC systems	421.6	78.26	−4
Natural ventilation	421.6	37.31	−54

9.2.4 *Summary*

A comparison of the four different systems and the base case, in terms of the carbon footprint, is shown below.

The comparison includes air-based chiller systems, but since it is not an effective low-carbon solution, it is not elaborated here. The table below summarizes the other alternatives for operational emission reduction.

According to the IGBC limit described in Chap. 2, Radiant cooling is within the comfort limit and releases the least carbon emissions (Fig. 9.5), (Table. 9.2).

The various iterations in line with the guidelines described in Chap. 5, have been simulated and the emissions quantified. These low-carbon alternatives provide a range of varying results which will be aggregated and discussed.

9.3 Summary

The design iterations applied to the baseline building as described in Fig. 9.1, indicate the following:

- **HVAC system iterations**—large carbon reductions of over 50% are possible with changes in the HVAC system. As one of the most significant energy users, it has the maximum impact on the operational emissions of a building. The embodied impact has not been assessed in this study as the quantity of smaterial is negligible, but it should be equally addressed and balanced.

- The refrigeration and air conditioning sector has also made a lot of progress using energy-efficient installations and new-generation fluids to reduce environmentally harmful emissions. Despite this huge progress, HVAC remains one of the most impactful areas in terms of carbon emissions, as they are required in many cases to achieve thermal comfort.
- The exploration of nature-based solutions through passive design or hybrid solutions like radiant cooling and taking advantage of advanced technology incorporation can yield significant reductions that would benefit the environment as well as overall comfort.

Chapter 10
The Potential of Carbon Assessments for Buildings

It is established that the manufacturing and building sectors form a large share of India's GHG emissions. Current building codes in India include the National Building Code (NBC), the Energy Conservation Act of 2002, as well as the Energy Conservation Building Code (ECBC) in 2007, and the Eco-Niwas Samhita (ENS) in 2018. All the codes significantly address operational carbon and energy, with minimal emphasis on embodied emissions [1]. Studies indicate that under current GHG policies, India's carbon emissions are on an upward trajectory until 2040 and would need additional mitigation strategies to reduce carbon emissions. [2]. Further, as discussed in Chap. 1, the urban population in India is projected to rise exponentially by 2050, and 70–80% of the urban infrastructure that will exist in 2050 is yet to be built [3].

We are currently facing a climate emergency, as highlighted by the 2018 UN Intergovernmental Panel on Climate Change (IPCC) report, depicting a stark contrast between a world with a 1.5 and 2 °C temperature increase. The latter scenario could lead to catastrophic climate breakdown, destroying entire ecosystems. The economic toll of heightened heating and cooling demand is predicted to quadruple by the century's end, with some consequences being irreversible. Urgent action is imperative to transform unsustainable consumption models.

The built environment sector, responsible for 39% of global carbon emissions, plays a crucial role in addressing the climate emergency. Decarbonizing this sector is among the most cost-effective strategies to mitigate climate breakdown. In 2018, aligning with the Paris Agreement, the World Green Building Council (WorldGBC) launched the Net Zero Carbon Buildings Commitment. This initiative aims to inspire climate leadership, urging businesses and governments to achieve net-zero operational carbon at both individual and mass scales. However, it emphasizes that operational carbon emissions are just one aspect of the broader story.

According to the Paris Agreement, in order to avoid an irreversible and catastrophic climate change trajectory, average global temperatures must not rise more than 2 °C. Since emissions accumulate in the atmosphere and there is limited time remaining before the arrival of this tipping point, emissions released now are more

S. Mohan and Vijayalaxmi J., *Embodied and Operational Carbon in Buildings*,
SpringerBriefs in Architectural Design and Technology,
https://doi.org/10.1007/978-981-97-7187-5_10

critical than emissions released later. The majority of a building's total embodied carbon is released upfront in the product stage at the beginning of a building's life. Unlike with operational carbon, there is no chance to decrease embodied carbon with updates in efficiency after the building is constructed. There is an urgent need to address embodied carbon now to meet short-term and long-term climate targets.

As the global population nears 10 billion, the building stock is projected to double, demanding a transformation in our sector to avert environmental crises. Failure to do so will risk doubling raw material consumption by mid-century, escalating emissions, and impacting climate. Carbon emissions, spanning manufacturing, transportation, construction, and end-of-life phases, contribute around 11% of global emissions—commonly termed embodied carbon. Often overlooked, these emissions significantly impact the environment. Upfront carbon emissions, occurring before a building is utilized, will account for half the carbon footprint of new constructions by 2050, jeopardizing a substantial portion of our remaining carbon budget. Urgent action is essential to mitigate these concerning trends.

This indicates an enormous potential to address carbon emissions in the early stage of urban development.

Broadly, there are a few strategies for reducing embodied carbon in buildings.

- Build less and reuse more by extending the life of existing buildings and materials.
- Build lighter and smarter with less of a given material (or floor area) to do the same work.
- Substitute low-carbon materials for high-carbon ones.
- Procure low(er)-carbon product selection.
- Use Carbon Sequestered building materials.

There are numerous ways to address carbon emissions, particularly for buildings in the current urban scenario of India. A few of the most effective measures are:

- Carbon reduction measures in regulations and policies. This can be done in the form of a quantifiable cap or target for building emissions.
- Standardization of tools, methodologies, and data sources to calculate life cycle carbon emissions, particularly for the early design stage.

However, setting regulations involves important and key stakeholder decisions, and may require time. A few initiatives such as an amendment to the Energy Conservation Act in 2022, to equip India with a carbon credit trading scheme are being drafted into an organizational framework [2].

Another volitional scheme for trading carbon credits at a competitive process, International Carbon Exchange Pvt.Ltd (ICX) was initiated in December 2022 [2].

Apart from these, more voluntary initiatives, such as green building rating systems and other frameworks, can be tailored and made more stringent to target carbon emissions.

These rating systems can also pave the way for broader regulations that can impact all construction projects.

This book addresses current green building rating systems (GBRS) and describes the potential impact of building design on carbon emissions.

These design decisions can be considered as guidelines in the scale of designers, as well as for green rating systems development, and their respective emissions calculated are summarized in Fig. 9.1. Table 10.1 summarizes the key strategies studied and their potential impact discussed in Chaps. 6–8. It can drive progress toward the achievement of net zero targets by increasing the ambition of green certifications. Further, green building certifications impact affluent stakeholders and could contribute to accelerating market advancement, workforce, and demand for low-carbon materials (Fig. 10.1).

10.1 Inferences and Recommendations

Key strategies to leverage embodied and operational carbon emissions include:

1. It is evident that the structural system materials and alternatives predominantly affect embodied carbon to a scale of up to 50% reductions. Structural systems are a parameter that cannot be changed after construction and need to be addressed consciously from the lens of carbon reductions at a design stage.
2. Current green building rating systems in India address materials and resources collectively under its credit criteria. Therefore, more focus on structure-specific reductions and criteria independently will lead to further integration of structure and design.
3. For single-layer envelope materials, natural materials like straw-based unfired earth brick and rammed earth perform better than AAC block and include up to 15% reduction over the conventional case. This highlights the necessity to use quantifiable low-carbon or carbon sequestering materials and material selection must be concurrent with material LCAs and targets.
4. For double-layer envelope materials with cavities, operational emission is impacted as well as embodied emission. It is important to understand the balance of embodied and operational emissions in a proposed building to propose efficient strategies. The code can address trade-offs between embodied and operational carbon to facilitate overall reduction. For example: the utilization of solar panels, decreases operational emissions but has high embedded emissions.
5. The use of alternative materials to replace carbon hotspots can impact more reductions, compared to replacing other materials. For example: In Chap. 4, effective carbon reductions were affected by replacing Aluminum frames and structures with alternative materials, though Aluminum forms less than 8% of the volume of materials used.

   Green building rating systems can address targeted strategies considering the most contributing materials to carbon emissions. It would require a certain degree of flexibility in targets and applicability.
6. It is seen that a large impact on operational emissions is due to systems and equipment. HVAC system changes can result in up to a 60% reduction in operational carbon emissions. However, the possibility of addressing these strategies

Table 10.1 Summary of results from a study in Chaps. 6–8

Parameter	Embodied carbon ($kgCO_2e/m^2$)	Operational carbon ($kgCO_2e/m^2/yr$)	Embodied carbon ($kgCO_2e/m^2/yr$) considering a life of 60 years	Whole life carbon ($kgCO_2e/m^2/yr$)
Base case	421.6	73.15	7.02	80.17
Conventional case	476	81	7.93	88.93
Type of structural material				
Precast concrete	453	73.15	7.60	80.75
Rice husk ash concrete	426	73.15	7.12	80.27
Bamboo leaf ash concrete	425	73.15	7.09	80.24
GGBS concrete	419	73.15	6.98	80.13
Type of structural system				
Bamboo reinforcement	350	73.15	5.83	78.98
Recycled steel	660	73.3	11.01	84.31
Alternative envelope materials				
CSEB	408.7	79.5	6.64	86.14
Rammed earth	398.6	79.4	6.81	86.21
Straw-based brick	390.2	72.98	6.50	79.48
AAC block	409.6	73.9	6.83	80.73
FaLG	397.2	79.4	6.62	86.02
FaLG with cavity	425	72.9	7.08	79.98
Porotherm	396	81.2	6.61	87.81
Alternative materials				
Filler slab	366.56	72.1	7.60	79.70
Bamboo shading	349	73.15	7.60	80.75
Systems configuration				
HVAC with VRF	421.6	50.81	7.60	58.41
Radiant cooling	421.6	44.55	7.60	52.15
Air-based HVAC systems	421.6	78.26	7.60	85.86
Natural ventilation	421.6	37.31	7.60	44.91

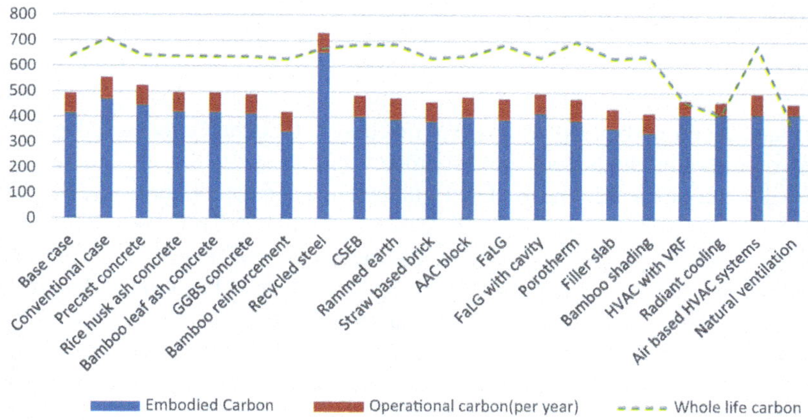

Fig. 10.1 Chart showing carbon data of the alternatives

at a design stage must be considered and addressed in rating systems, to reduce emissions further.

On a broader scale, some strategies for the rating system development include:

- Introduce or mandate whole life carbon assessment for embodied and operational carbon. Setting a standard methodology, database, and scope for comparability and reduction.
- Setting a range of targets for embodied carbon emissions of materials and addressing trade-offs, for various building types for providing reference and direction.
- Increase weightage for the use of green building materials. This not only aids the building industry but also the manufacturing sector, which is responsible for a large share of carbon emissions.
- Possibility to revise scoring criteria to focus on emission reduction as a key goal, particularly in India.

The IGBC Net Zero Carbon rating system is one such attempt to address whole life carbon emissions in India, effected by the Indian Green Building Council. However, it is not as wholistic as the studied rating systems because it delineates targets without addressing the associated strategies. The targets for this system are:

- Achieve embodied carbon of less than or equal to 700 kg CO_2e/square meter benchmark (and) offset at least 90% of the total operational carbon.

However, the intention and carbon consciousness should be integrated into other rating systems as well.

10.2 Further Discussion

One of the key outcomes of Chaps. 6, 7, and 8, lies in establishing comparability of quantified carbon emissions. As discussed, this can be achieved with a standardization, of methods, tools, and sources.

This would require a baseline model or a target emission, which forms the basis of the criteria in green rating systems. In the case of a baseline comparison, unless the material and specification are established, the method would be less stringent than the target setting. It would also present a series of problems related to the definition of conventional, which varies based on geography and design.

In the case of target setting, however, the question of the metrics used and further granularity of how to differentiate embodied and operational emissions becomes the predominant challenge.

In Table 10.1, it is quantified based on whole-life carbon ($kgCO_2e/m^2/year$). Though customarily, it includes the whole life cycle, in this study, it accounts for A1–A3 and B6, which form more than 70% of the life cycle emissions. This boundary includes embodied and operational emissions, integrated into one quantifiable metric.

While this is effective for comparison, it is not an accurate measure of the emissions and may lead to inefficient assumptions. The challenges in the use of this metric for target setting are described below.

- As this is divided across the life cycle of the building, the variability of emission factors over the period must be accounted for. While operational carbon can account for carbon factors based on projected electricity mixes, embodied carbon factors are more specific and account for only current carbon factors.

  This may result in the setting of less ambitious carbon targets in the utilization of this metric.
- Further, over the lifecycle, it is recognized that operational emissions account for a major share of emissions (more than 70%). The aggregated metric also indicates this apportionment. Therefore, impacting operational emissions would reduce the metric more than embodied emissions. This is also indicated in Table 10.1.

The balance and sensitivity to embodied emissions are therefore diluted in this approach.

Despite the challenges, there is a necessity to provide a unified target to address carbon emissions. The benefits of a quantifiable target are discussed below.

- It would ease carbon-specific target setting and create an easier path to address carbon emissions in rating systems and standards. Particularly, it would visualize a goal and help stakeholders and investors better understand the final target.
- In the early design phase of buildings, it could allow easier comparisons and facilitate integrations with other metrics like radiation, daylight, and energy.

Therefore, while a target is necessary, it may or may not be aggregated into a single comprehensive value. Further review and comparison are required to establish a relevant and comprehensive metric of measurement for target setting. For example:

While Denmark uses a single metric $kgCO_2/m^2$/ year, France and Sweden use carbon emission intensity (kg CO_2/m^2) and the Netherlands uses a monetary-based metric of EUR/m_2/year [4].

It is hence acknowledged that there has been a consistent effort in bringing carbon metrics and methodology of carbon assessment into several national and international standards. Each geographical context has its own challenges based on location, policy, social and economic landscape, demographics, and others.

For developing countries like India, which is also a tropical country, global warming will have a critical impact and hence addressing carbon becomes more imperative. It also poses a huge economic challenge, as it is estimated that to achieve the 2070 net zero target, India needs an investment of $10.1 trillion beginning now [5]. While the challenges are unique and very real, the solution must be flexible enough to adapt to the relevant context and rigorous enough to effectively reduce carbon emissions. For example, IGBC Net Zero carbon sets targets for embodied carbon using carbon emissions intensity (kg CO_2/m^2) and operational carbon as $kgCO_2/m^2$/ year. While this addresses both types of carbon emissions with the relevant metric, it does not differentiate the type of building or climatic context.

10.3 Potential for Future Study

- *Embodied Carbon*: As discussed in Chap. 2, the majority of planning policies in India for the real estate and construction sector are based on the operational phase. With the forecasted new construction, embodied carbon needs to be given importance and regulated in India through policy intervention.
- *Climate Impact*: Further, the ECBC and ENS address emissions through energy (EPI/EUI) and not directly through carbon emissions. While reducing energy consumption is key to carbon reduction, a metric is necessary to understand climate impact and operational phase impact.

Figure 10.2 describes some of the key challenges faced and the areas of improvement for the Indian construction sector.

- *Standardized tools*: The use of software tools is also a key area of interest. The green-rated buildings reviewed in Chap. 2, use Embodied carbon calculator (EC3) which uses EPD data for emission factors, and Carbon Designer/One Click LCA to determine embodied carbon emissions. The results were variable by 10% with the same assumptions and specifications. A study of the available tools and a review of a suitable one for the Indian context may simplify the LCA study and the accessibility of such tools.
- *Environmental product declarations*: Another key area of development is the EPD data, particularly in the Indian manufacturing sector. The more the availability of EPDs, the accuracy of the emission factor used can be improved. However, it is an intensive process with the LCA study by tracking emissions, compilation,

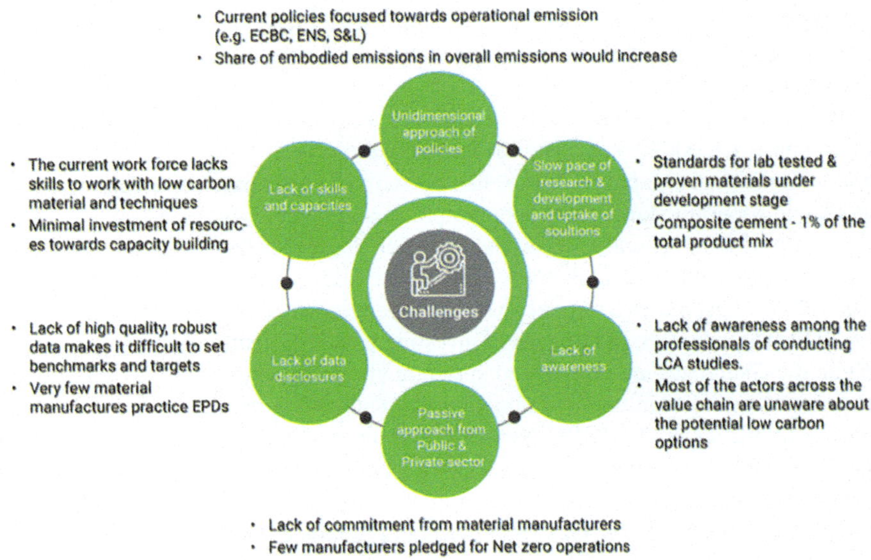

Fig. 10.2 Key challenges for decarbonization of the building sector in India [6]

verification by a third party, registration, and publication. This disclosure needs to be encouraged to provide reliable and consistent carbon factor data for future studies.

- *Decarbonize allied sectors*: Carbon emissions are also a result of all phases of construction. Therefore, there is a necessity to decarbonize the wider scale sectors like manufacturing or transport.
- *National standardized database*: There is a need for a national database and reporting platform for carbon emissions of the built environment in India. Lack of accessibility to data and technology is one of the major barriers to carbon assessments.

The route to understanding the carbon emissions of the built environment is complex and varied, with many branches. However, it is clear that for any building, carbon emissions must be considered right from the conceptual stage across the life of the building. It can be done through various methods like design or utilizing rating systems as well as local practices. However, the essence remains the same, which is the integration of carbon reduction practices. Quantification of these practices and target setting gives a reference and a direction for further decarbonization of the building sector to meet the country's ambitious net zero goals.

While high-level strategies (like material efficiency in designing structural systems or using lower carbon building materials and assemblies) are clear, available research does not tie these strategies to reliable percentage reductions across buildings. Once building LCA has been more broadly adopted and more LCA results are available, analysis of the pool of LCA results should be able to fill this research gap. However, even as prescriptive design strategies become available, WBLCA will always have a role in assessing novel materials and design solutions.

References

1. Khan S, Bhushan S (2023) India's building code has a blind spot for a whole category of emissions, April 13. Retrieved from Science (The Wire): https://science.thewire.in/environment/india-building-code-embodied-emissions/
2. Chateau J, Dang G, MacDonald M, Spray J, Thube S (2018) A framework for climate change mitigation in India. Working Paper (International Monetary Fund)
3. Urbanet (2018) Infographics: urbanisation and urban development in India, July 31. Retrieved from Urbanet: https://www.urbanet.info/urbanisation-in-india-infographics/
4. Steinmann J, Rock M, Lutzkendorf T, Allacker K, Le Dan X (2022) Whole life carbon models for the EU27 to bring down embodied carbon emissions of buildings. Ramboll, Belgium
5. Asia Society Policy Institute (2022) Getting India to net zero. Asia Society Policy Institute
6. Bhardwaj S, Tewari D (2022) Tackling embodied carbon from India's building sector, March 28. Retrieved from Alliance for an Energy Efficient Economy: https://aeee.in/tackling-embodied-carbon-from-indias-building-sector/

Uncited References

7. Blesson S, Rao A (2023) Agro-industrial-based wastes as supplementary cementitious or alkali-activated binder material: a comprehensive review. Innovative Infrastructure Solutions
8. Cang Y, Yang L, Luo Z, Zhang N (2020) Prediction of embodied carbon emissions from residential buildings with different structural forms. Sustainable Cities and Society
9. Central Electricity Authority, Ministry of Power (2022) CO_2 Baseline database for the Indian Power Sector. Government of India, Ministry of Power, New Delhi
10. Chen X, Yang H, Lu L (2015) A comprehensive review on passive design approaches in green building rating tools. Renew Sustain Energy Rev 1425–1436
11. Diaz AV, Lopez AF, Bugallo PB (2022) Analysis of biowaste-based materials in the construction sector: evaluation of thermal behaviour and life cycle assessment (LCA). Waste Biomass Valorization 4983–5004
12. Esau R, Jungclaus M, Olgyay V, Rempher A (2021) Reducing embodied carbon in buildings—low cost and high value opportunities. RMI
13. Fnais A, Rezgui Y, Petri I, Beach T, Yeung J, Ghoroghi A, Kubicki S (2022) The application of life cycle assessment in buildings: challenges, and directions for future research. Int J Life Cycle Assess 627–654
14. GBPN (2022) Towards zero carbon: building policies in India 2022. GBPN
16. Gibbons O, Orr J (2022) How to calculate embodied carbon. Institute of Structural Engineers
17. Golisano Institute for Sustainability (2020) What is life cycle assessment (LCA)?, July 2. Retrieved from RIT educational blogs: https://www.rit.edu/sustainabilityinstitute/blog/what-life-cycle-assessment-lca#:~:text=An%20LCA%20is%20a%20systematic,that%20make%20up%20industrial%20production
18. Google Scholar (2023) Google scholar. Retrieved from https://scholar.google.co.in/scholar?hl=en&as_sdt=0%2C5&as_ylo=2023&q=energy+building&btnG=
19. GRIHA (n.d.) Case studies. Retrieved from GRIHA India: https://www.grihaindia.org/case-study

20. GRIHA Council (2021) GRIHA V.2019, 3rd ed. The Energy and Resources Institute (TERI), New Delhi
21. Hakkinen T, Kuittinen M, Ruuska A, Jung N (2015) Reducing embodied carbon during the design process of buildings. J Build Eng 1–13
22. ILFI (2024) Living Building Challenge 4.1—Program manual. ILFI
23. ILFI (2024) Zero carbon certification 1.1—Program manual. ILFI
24. Illankoon SI, Tam V, Le K, Shen L (2017) Key credit criteria among international green building rating tools. J Clean Prod 209–220
25. Indian Green Building Council (IGBC) (2016) IGBC green new buildings rating system. IGBC, Hyderabad
26. International Finance Corporation (IFC) (2017) India construction materials database of embodied energy and global warming potential. European Union and IFC
27. Junnila S, Horvath A (2003) Life-cycle environmental effects of an office building. J Infrastruct Syst 4
28. Kanaujia V, Reddy AS, Kumar K (2017) Comparative review of indian green building rating system. J Energy Res Environ Technol 194–198
29. Kjendseth M, Fufo SM, Andresen I (2018) Design strategies for low embodied carbon in building materials. In: Pomponi F, De Wolf C, Moncaster A (2018) Embodied carbon in buildings—measurement, management and mitigation. Springer, pp 323–339
30. Kurian R, Kulkarni KS, Ramani PV, Meena CS, Kumar A, Cozzolino R (2021) Estimation of carbon footprint of residential building in warm humid climate of India through BIM. Energies 14
31. Lahiri A, Dixit S (2015) Alternatives to cement in concrete—a review. Int J Sci Eng Res 8(8)
32. LEED v4 (2021) Reference guide for building design and construction. Retrieved from USGBC: https://www.usgbc.org/guide/bdc#credit
33. Li S, Yan H, Chen J, Shen L (2017) A life cycle analysis approach for embodied carbon for a residential building. In: Proceedings of the 20th international symposium on advancement of construction management and real estate. ResearchGate, pp 1185–1196
34. Luo Z, Yang L (2015) Embodied carbon emissions of office building: a case study of China's 78 office buildings. Build Environ 95(1)
35. Marsh R, Rasmussen NF, Birgisdottir H (2018) Embodied carbon tools for architects and clients early in the design process. In: Pomponi F, De Wolf C, Moncaster A (eds) Embodied carbon in buildings—mitigation, measures and management. Springer, pp 167–190
36. Maruthupandian G, Saravanan R, Kumar S, Sivakumar B (2016) A study on bamboo reinforced concrete slabs. J Chem Pharm Sci 978–980
37. Masson-Delmotte V, Zhai P, Pörtner HO, Roberts D, Skea J, Shukla PR … Waterfield T (2018) Global warming of 1.5°C. An IPCC special report on the impacts of global warming of 1.5°C above pre-industrial levels and…. IPCC
38. Ministry of External Affairs, GOI (2021) Media Center. Retrieved from National Statement by Prime Minister Shri Narendra Modi at COP26 Summit in Glasgow: https://www.mea.gov.in/Speeches-Statements.htm?dtl/34466/National+Statement+by+Prime+Minister+Shri+Narendra+Modi+at+COP26+Summit+in+Glasgow
39. Muralikrishna IV, Manickam V (2017) Chapter Five—Life cycle assessment. In: Muralikrishna IV, Manickam V (eds) Environmental management. Elsevier, pp 57–75
40. Nanz L, Rauch M, Honermann T, Auer T (2019) Impacts on the embodied energy of rammed earth facades during production and construction stages. J Facade Des Eng 7(1):75–88
41. Net Zero Energy Buildings (NZEB) (n.d.) Detailed case studies. Retrieved from NZEB: https://nzeb.in/case-studies/detailed-case-studies-2/
42. Olhoff A, Christensen J (2023) Emissions Gap Report 2023: broken record. UNEP, Nairobi
43. One Click LCA (2018) The embodied carbon review. One Click LCA
44. Pallas G (2021) Life cycle-based sustainability standards and guidelines, May 11. Retrieved from Pre-sustainability: https://pre-sustainability.com/articles/lca-standards-and-guidelines/
45. Pan W, Teng Y (2021) A systematic investigation into the methodological variables of embodied carbon assessment of buildings. Renew Sustain Energy Rev 141

46. Passe U, Battaglia F (2015) Designing spaces for natural ventilation—an architect's guide. Routledge
47. Pomponi F, Moncaster A (2018) Scrutinising embodied carbon in buildings: the next performance gap made manifest. Renew Sustain Energy Rev 2431–2442
48. Praseeda K, Reddy B, Mani M (2015) Embodied energy assessment of building materials in India using process and input–output analysis. Energy Build 677–686
49. Prideaux F, Allacker K, Crawford RH, Stephan A (2024) Integrating life cycle assessment into the building design process—a review. In: Environmental research—infrastructure and sustainability
50. Puri HS (2019) How India is building cities of the future. The Economic Times, June 2019
51. RICS (2012) Methodology to calculate embodied carbon of materials. RICS
52. Smith MR (2015) "Green" building in India: a comparative and spatial analysis of the LEED-India and GRIHA rating systems. Asian Geogr 73–84
53. The Energy and Resources Institute (TERI) (2024) TERI's solutions for sustainable development. TERI, New Delhi
54. Times of India (2023) Air conditioning in India, June 22. Retrieved from Times of India: https://timesofindia.indiatimes.com/home/environment/air-conditioning-in-india-poses-risk-of-dramatic-rise-in-emissions-study/articleshow/101190767.cms
55. UGreen (2023) EDGE certification—the complete guide, December 25. Retrieved from UGreen: https://ugreen.io/edge-certification-the-complete-guide/#:~:text=Scope%3A%20EDGE%20Certification%20is%20focused,buildings%2C%20interiors%2C%20and%20neighborhoods.
56. UNFCCC (2018) Key aspects of the Paris Agreement. Retrieved from United Nations Climate Change: https://unfccc.int/most-requested/key-aspects-of-the-paris-agreement#:~:text=The%20Paris%20Agreement's%20central%20aim,further%20to%201.5%20degrees%20Celsius.
57. United Nations (2014) World urbanization prospects. Department of Economic and Social Affairs. United Nations
58. United Nations Environment Programme (2024) Global status report for buildings and construction. UNEP, Nairobi
59. Vijayalaxmi J (2009) The thermal comfort of a naturally ventilated house resulting from the evaporative cooling of a ceiling fan in the hot-humid climate of Chennai. Int J Vent 135–144
60. Weir M, Rempher A, Esau R (2023) Embodied carbon 101: building materials, March 27. Retrieved from RMI: https://rmi.org/embodied-carbon-101/
61. World Green Building Council. (2019). *Bringing Embodied Carbon Upfront*. Ramboll and World GBC.
62. Zuiderveen E, Kuipers K, Caldeira C, Hanssen SV, van der Hulst M, de Jonge MM, … Huijbregts M (2023) The potential of emerging bio-based products to reduce environmental impacts. Nat Commun 14

SPRINGER NATURE

GPSR Compliance

The European Union's (EU) General Product Safety Regulation (GPSR) is a set of rules that requires consumer products to be safe and our obligations to ensure this.

If you have any concerns about our products, you can contact us on ProductSafety@springernature.com

In case Publisher is established outside the EU, the EU authorized representative is:

Springer Nature Customer Service Center GmbH
Europaplatz 3
69115 Heidelberg, Germany

The manufacturer's authorised representative in the EU is Springer Nature Customer Service Centre GmbH, Europaplatz 3, 69115 Heidelberg, Germany. If you have any concerns regarding our products, please contact ProductSafety@springernature.com

Printed and bound by CPI Group (UK) Ltd, Croydon, CR0 4YY

30/09/2026

02225267-0004